COLOUR G
Microbiolo

T.J.J. Inglis DM MRCPath DTM&H

Department of Microbiology,
University of Leeds,
Leeds, West Yorkshire, UK

A.P. West BSc FIMLS

Department of Microbiology,
University of Leeds,
Leeds, West Yorkshire, UK

Churchill Livingstone

EDINBURGH LONDON MADRID MELBOURNE NEW YORK AND TOKYO 1993

CHURCHILL LIVINGSTONE
Medical Division of Longman Group UK Limited

Distributed in the United States of America by
Churchill Livingstone Inc., 650 Avenue of the Americas,
New York, N.Y. 10011, and by associated companies,
branches and representatives throughout the world.

© Longman Group UK Limited 1993

All rights reserved. No part of this publication may be
reproduced, stored in a retrieval system, or transmitted in any
form or by any means, electronic, mechanical, photocopying,
recording or otherwise, without either the prior written
permission of the publishers (Churchill Livingstone, Robert
Stevenson House, 1–3 Baxter's Place, Leith Walk, Edinburgh
EH1 3AF), or a licence permitting restricted copying in the
United Kingdom issued by the Copyright Licensing Agency
Ltd, 90 Tottenham Court Road, London W1P 9HE.

First published 1993

ISBN 0443-03972-0

British Library Cataloguing in Publication Data
A catalogue record for this book is available from the British
Library.

Library of Congress Cataloging in Publication Data
A catalogue record for this book is available from the Library
of Congress.

Publisher
Timothy Horne
Project Editor
Jim Killgore
Production
Nancy Henry
Designer
Design Resources Unit
Sales Promotion Executive
Marion Pollock

The publisher's policy is to use **paper manufactured from sustainable forests**

Printed in Hong Kong
LYP/01

Preface

First contact with clinical microbiology for most people is through the diagnosis of infectious diseases. Diagnostic microbiology provides a rational basis for the treatment of infection and for the maintenance of a safe hospital environment. It includes the sub-specialties of virology, mycology and parasitology and incorporates elements of immunology and molecular genetics. In recent years, the discovery of new agents of infection and improvements in laboratory technology have made the discipline increasingly complex.

This book aims to open the doors on the clinical microbiology laboratory, so that the reader can understand the complex sequence of events that begins and ends with the patient. The book is not an exhaustive catalogue of procedures, but instead places an emphasis on the common and the rare-but-important. The largely visual nature of microbiology has been employed to make the subject as accessible as possible. This text is therefore suited both to medical undergraduates and to clinicians preparing for postgraduate exams.

We are grateful to many colleagues and friends for their assistance with the preparation of this book, in particular: Mr R. Foster, Mycology Laboratory, Department of Microbiology, University of Leeds; Mr S. Toms, Department of Pathology, University of Leeds; Mr J. O'Neill, Virology Department, Public Health Laboratory, Leeds; Mr P. Parnell, Department of Microbiology, Leeds General Infirmary; and Dr K. Kerr, Department of Microbiology, Leeds General Infirmary who reviewed the manuscript.

1993

T.J.J.I.
A.P.W.

Contents

1. Specimens—sampling and transport — 1
2. Microscopy — 3
3. Culture — 9
4. Benchtop tests — 17
5. Confirmatory tests — 19
6. Antibiotics in the laboratory — 21
7. Urinary tract infection — 25
8. Sexually transmitted diseases — 35
9. Gastrointestinal tract infections — 47
10. Respiratory tract infections — 57
11. Meningitis — 67
12. Soft tissue infections — 79
13. Septicaemia — 89
14. Laboratory methods in infection control — 99
15. Virology — 103
16. Mycology — 113
17. Parasitology — 121

Index — 133

1 / Specimens—sampling and transport

Diagnostic microbiology is a process that begins and ends with the patient. In between, a range of laboratory tests and procedures have to be chosen. The specimen and its request form are the initial point of contact between clinical practice and the diagnostic laboratory.

Sampling Important considerations at the stage of specimen collection include:
- *What to send?*
  - From what body site?
  - From other sites?
- *How to collect it?*
  - How to prevent contamination?
  - What container?
  - How much to send?
- *What clinical data might be useful* (Fig. 1)?
- *Urgency*
  - Will results alter management now?
  - Will organisms die in prolonged transit?
- *Safety*
  - Is the container properly sealed?
  - Does it require special hazard label (Fig. 2)?

Transport Prolonged transport may cause deterioration of most types of clinical specimen. In some cases a very short transit time is required to ensure survival of causative organisms. Deterioration of specimens is due to:
- death of fastidious species
- overgrowth of minority species
- environmental contamination
- dessication.

Special transport media are required for viruses and *Chlamydia* spp.

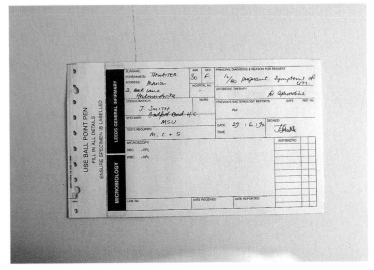

Fig. 1 Specimen request form complete with relevant clinical information.

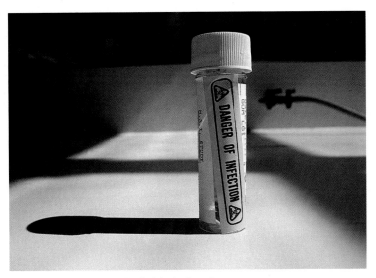

Fig. 2 Specimen container labelled for infection hazard.

2 / Microscopy

Microscopic techniques are of fundamental importance to microbiology since bacteria, viruses, fungi and most parasites are not easily visible to the naked eye. The following light and electron microscopic techniques are used:

- *Light microscope*
  - white light
  - ultraviolet
  - dark ground
  - phase contrast.
- *Electron microscope*
  - transmission
  - scanning.

With the sole exception of viruses, all the major classes of microorganisms are visible under the light microscope. Samples can be prepared in a variety of ways to make organisms more easily visible, and to demonstrate structural features.

Wet films

One of the simplest microscopic techniques. Organisms are suspended in a suitable solution (e.g. saline), without stain.

Uses
- Bacterial motility.
- Leucocyte count (e.g. CSF, urine) (Fig. 3).
- Intestinal parasites.
- Vaginal *Trichomonas* sp.
- Dermatophytic fungi.

Single step stains

Procedures using only one dye or stain are usually limited by stain properties to specific applications.

Stains
- Capsular stain—detection of bacterial capsules by 'negative' stain (Fig. 4).
- Acridine orange—detection of bacteria under ultraviolet light (Fig. 5).
- Calcofluor white—examination of fungi.

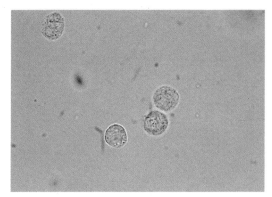

Fig. 3 Unstained leucocytes in suspension.

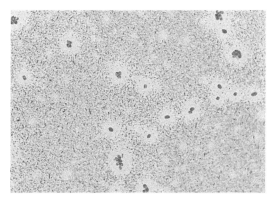

Fig. 4 Bacterial capsules, apparent as 'negative' staining haloes around bacilli.

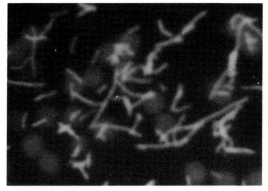

Fig. 5 Acridine-orange-stained bacilli visible under ultraviolet light.

Gram stain

Used extensively to determine the shape and cell-wall properties of medically important bacteria.

Method The method includes the following principal stages:
1. bacteria fixed to glass slide by heating
2. methyl violet (initial stain)
3. iodine solution as mordant
4. acetone or ethyl alcohol decolourant
5. carbol fuchsin (counterstain).

Rationale The appearance of bacteria stained by the Gram method is determined by the cell-wall structure. In Gram-positive bacteria, a thick layer of peptidoglycan retains the methyl violet dye during decolorization, making the organisms appear purple (Fig. 6). In Gram-negative bacteria the cell wall contains a much thinner peptidoglycan layer which does not retain methyl violet. Gram-negative bacteria therefore have the pink colouring of the counterstain (Fig. 6).

The shape of Gram-stained bacteria also helps to classify and identify them (Fig. 7).

Uses Principal applications include examination of:
- bacteria already in pure culture
- material from clinical specimens.

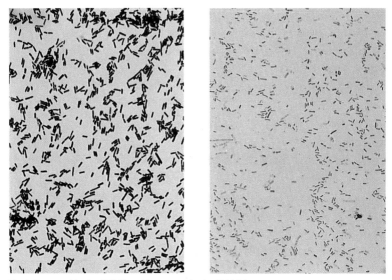

Fig. 6 Gram-positive bacilli (left) and Gram-negative bacilli (right).

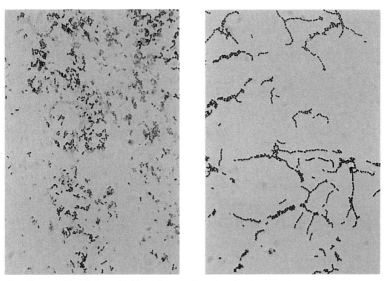

Fig. 7 Gram-positive diplococci (left) and Gram-positive chain forming cocci (streptococci) (right).

Ziehl–Neelsen stain

This stain is used to detect acid-fast bacteria. These bacteria possess mycolic acid and other substances in their cell walls that make them resistant to staining by simpler methods.

Method The method differs substantially from the Gram stain and involves the following stages:
1. heat fix
2. carbol fuchsin driven into organisms by heating
3. acid and alcohol decolourisation
4. counterstain.

Uses The principal uses include detection of:
- mycobacteria in clinical specimens (Fig. 8)
- mycobacteria in pure culture on/in laboratory media
- other acid-fast bacteria (e.g. *Nocardia* spp.)—requires modification (Fig. 9).

An alternative staining method uses auramine instead of concentrated carbol fuchsin. The stained preparation is viewed under ultraviolet light.

Spore stain

Some bacteria survive adverse conditions by producing spores. The shape and position of spores in the organism may help in its identification. Since spores are resistant to Gram stain, they are stained by a modified Ziehl–Neelsen procedure (Fig. 10).

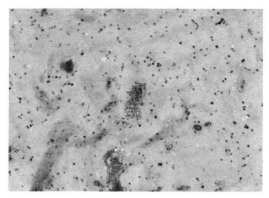

Fig. 8 Mycobacteria in sputum smear from patient with tuberculosis.

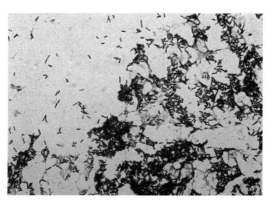

Fig. 9 *Nocardia* sp. stained by modified Z.N. procedure.

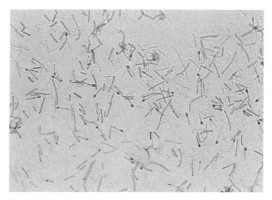

Fig. 10 *Clostridium tetani* with typical drumstick, terminal spores.

3 / Culture

Further useful information can be obtained about organisms by cultivating them under artificial conditions. To do this, bacteria must be exposed to the optimum temperature and gaseous atmosphere, and have access to the appropriate metabolic substrates. Substrates are provided by a range of solid and liquid media, some of which are described below.

Solid media

Non-selective Commonly used examples are blood and chocolate (haemolysed blood) agars, both of which are rich in nutrients and free from added inhibitory substances (Fig. 11). Both types of agar will support the growth of many different species of medically important bacteria, including some more fastidious species.

Solid media are inoculated by spreading material over a small area on the surface (the 'well'), and then spreading this out over the remainder of the plate with a sterile wire loop so that the density of the inoculum falls with each set of streaks (Fig. 12). This technique aims to provide growth in discrete, single colonies at some point on the plate (Fig. 13).

The following information can be obtained by examination of bacterial growth on the surface of non-selective media and then used in the preliminary identification of organisms.

- *Colony:* size, shape, colour, pigmentation.
- *Haemolysis:*
  α—green discolouration around colony
  β—clearing of blood agar around colony.
- *Smell*—may be characteristic of a species.

Fig. 11 Blood and chocolate (haemolysed blood) agars.

Fig. 12 Inoculation of agar plate with wire loop.

Fig. 13 Agar plate after incubation, showing single colony growth at end of inoculated area.

Selective (solid media)

Non-selective solid media are unsuitable for isolating a single bacterial species from specimens that are prone to contamination with large numbers of commensal organisms. A large range of selective media has been developed to help isolation of medically important species in the presence of commensal species. Key ingredients of selective media include:

- preferred substrates
- inhibitor substances, e.g. antibiotics
- indicator systems, to aid detection
- buffers.

Commonly used examples are:

- *MacConkey agar:* contains lactose and indicator. It demonstrates lactose fermenting/nonfermenting Gram-negatives (Fig, 14).
- *CLED (cysteine, lactose, electrolyte-deficient) agar:* also shows lactose fermentors. It is used for urine culture (Fig. 15).
- *XLD (xylose, lysine decarboxylase) agar:* includes H_2S indicator system. It is used for isolating salmonellas from faeces (Fig. 16).
- *VCAT (vancomycin, colistin, amphotericin, trimethoprim) agar*: used to grow *Neisseria gonorrhoeae* from genitourinary specimens (Fig. 17).

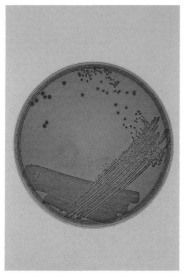

Fig. 14 MacConkey agar: lactose fermenters appear pink.

Fig. 15 CLED agar: lactose fermenters appear yellow.

Fig. 16 XLD agar: H_2S producers appear black.

Fig. 17 VCAT agar with growth of *N. gonorrhoeae*.

Liquid media

It is sometimes necessary to grow organisms in liquid media which lack the solidifying agar component. It is not possible to observe the characteristics of single colony growth, and the Gram stain appearance of bacteria from liquid culture may differ from identical organisms grown on solid media.

Uses Uses for liquid media include:
- *Specimen inoculation medium:* e.g. blood culture, where the patient's blood is added to bottles containing liquid media, such as brain–heart infusion or fastidious anaerobe broth (Fig. 18).
- *Recovery of fastidious organisms:* e.g. Kirschner's broth for mycobacteria, or Robertson's cooked meat broth for anaerobic bacteria (Fig. 19).
- *Metabolic reactions:* e.g. urease broth (contains urea and indicator) (Fig. 20), peptone water.
- *Antibiotic susceptibility testing:* e.g. minimum inhibitory concentration, performed in a series of broth dilutions of antibiotic.

Fig. 18 Bacterial growth in blood culture bottle.

Fig. 19 Anaerobic growth in Robertson's cooked meat broth.

Fig. 20 Positive and negative reactions in urease broth.

Incubation

This provides suitable temperature and atmospheric conditions for microbial growth. Most medically important bacteria will grow at or around 37°C, which is the commonest temperature used. Important exceptions to this are:

- 43°C—*Campylobacter* sp.
- 25°C—*Listeria* sp. motility test.

Bacteria differ in their atmospheric requirements for growth:

- *aerobic:* grow in air
- *anaerobic:* no growth in presence of oxygen
- *facultative anaerobe:* will grow in aerobic or anaerobic conditions
- *capnophilic:* require carbon dioxide
- *microaerophilic:* require strictly limited oxygen.

Apparatus Various systems are used to provide these temperature and atmospheric requirements, including:

- thermostatically controlled incubator
- CO_2 incubator (Fig. 21)
- gas jar, with gases supplied by an external source (cylinder) or an activated sachet (Fig. 22)
- anaerobic cabinet (Fig. 23).

Fig. 21 CO_2-enriched air incubator.

Fig. 22 Gas jar with activated sachet.

Fig. 23 Anaerobic cabinet.

4 / Benchtop tests

These are simple, rapidly performed tests used in the preliminary identification of bacteria, which include catalase, oxidase and coagulase reactions. Controls are required in each case.

Catalase test — Hydrogen peroxide is added to a sample colony. Oxygen bubbles form rapidly in a positive test (Fig. 24).

It is used to distinguish staphylococci (positive) from streptococci (negative).

Oxidase test — Indophenol blue is produced when cytochrome oxidase in a sample colony reacts with tetramethyl phenylenediamine dihydrochloride (Fig. 25). It is used to distinguish *Enterobacteriaceae* (negative) from *Pseudomonas* sp. (positive). *Campylobacter* sp., *Neisseria* sp. and *Vibrio* sp. also give positive reactions.

Coagulase test — Plasma is added to a suspension of staphylococci to produce a coarse, grainy suspension when *S. aureus* is present (Fig. 26). This result may require confirmation with an overnight test-tube version of the test, or the DNAase test in which DNA incorporated in agar is lysed around colonies of *S. aureus*. Zones of clearing are made visible by addition of hydrochloric acid which precipitates the DNA.

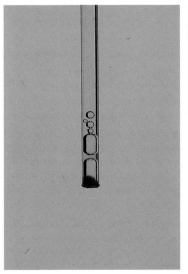

Fig. 24 Bubbles produced in catalase test by staphylococci.

Fig. 25 Positive oxidase test due to *Pseudomonas* sp.

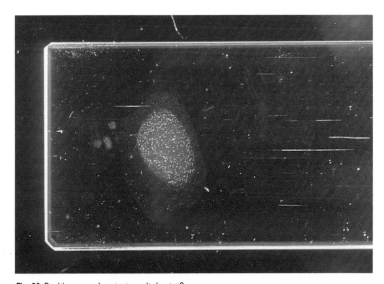

Fig. 26 Positive coagulase test result due to *S. aureus*.

5 / **Confirmatory tests**

These are used for more detailed identification of organisms. It may be necessary to issue a preliminary report when the time required to perform confirmatory tests might adversely affect clinical outcome.

Biochemical tests
Reactions in the presence of a variety of substrates and indicator substances are used to classify bacteria into genera and species. Tests may be performed by inoculation of organisms into single containers (see Fig. 20, p. 14), or into a gallery of reaction chambers (Fig. 27). Results of these tests can be interpreted with the help of taxonomic tables. Test results may indicate that organisms isolated from different sites have the same identity.

Immunological tests
The specificity of immunological reagents suits them to the identification of many medically important microorganisms. Agglutination reactions are most frequently used for reasons of convenience. Commonly used examples are:
- direct agglutination—salmonella and shigella serotyping (Fig. 28)
- latex agglutination—streptococcal grouping (Fig. 29)
- coagglutination—*N. gonorrhoeae* confirmation.

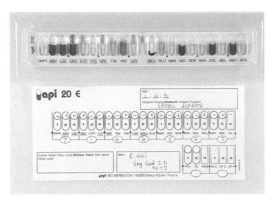

Fig. 27 Biochemical reactions performed in multiple reaction gallery.

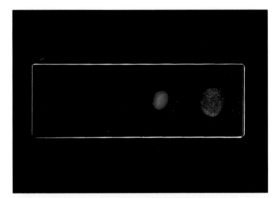

Fig. 28 Agglutination reaction with specific antiserum (+ve on right).

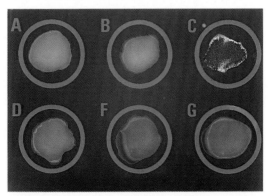

Fig. 29 Agglutination reaction with antibody-coated latex particles.

6 / Antibiotics in the laboratory

Antibiotic susceptibility testing

One of the most important functions of the diagnostic laboratory is to test the susceptibility of microbial pathogens to commonly used antibiotics. Susceptibility testing is particularly important when sensitivity to a given antibiotic agent cannot be reliably predicted.

Method The disc diffusion method is commonly used. It involves:

- spreading a suspension of test and control bacteria (of known susceptibility) over adjacent parts of an agar plate (Fig. 30)
- placing antibiotic containing discs on the agar surface at the interface between test and control (Fig. 31)
- incubating overnight.

Rationale Antibiotics diffuse into the agar and inhibit the growth of sensitive bacteria in a semicircular zone around the disc. When resistance to a given agent is present, the zone radius will be reduced or there will be no zone at all (Fig. 32).

Limitations In some cases, zone size is not a reliable indicator of susceptibility, for example:

- *S. aureus* may be resistant despite a large inhibition zone due to β lactamase.
- Some *Enterobacteriaceae* may have inducible chromosomal β lactamases. Inhibition zones to certain β lactam antibiotics may not be reduced despite production of this resistance factor.
- Large molecule antibiotics may produce small zones due to slow diffusion in agar.

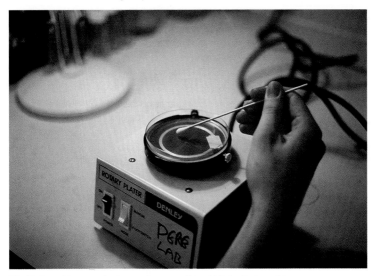

Fig. 30 Bacterial suspension is spread on agar plate with help of rotary plater.

Fig. 31 Antibiotic disc dispenser in use.

Fig. 32 Plate after incubation showing resistance of test organism to ampicillin (AMP disc).

Therapeutic drug monitoring

It may be necessary to measure the serum levels of an antibiotic during therapy. The principal reasons for measuring drug levels are:
- detection of toxic levels
- demonstration of subtherapeutic levels.

The most commonly measured agents are:
- aminoglycosides
- vancomycin
- chloramphenicol.

Methods A variety of methods are used to determine antibiotic concentrations in human body fluids (most often serum), and include two principal approaches:
- bioassay
- automated assay system.

It is essential that the laboratory is informed of any other antibiotics given to the patient in the previous 24 hours, as these may interfere with the assay system.

Bioassay
In the bioassay, wells are cut in an agar plate pre-seeded with an organism of known sensitivity to the antibiotic in question. Pre- and post-dose serum specimens are loaded into the wells along with a series of known control specimens, from where antibiotic diffuses into the agar to produce a zone of inhibition (Fig. 33). Zone diameters are used to calculate the antibiotic concentration in the samples.

Automated assay
Automated assays (Fig. 34) are not subject to the delay caused by incubation of the bioassay, and can produce results very rapidly.

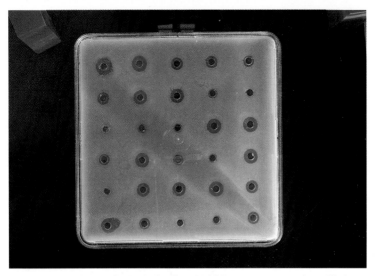

Fig. 33 Antibiotic bioassay after incubation. Growth inhibition zones correspond to antibiotic concentration.

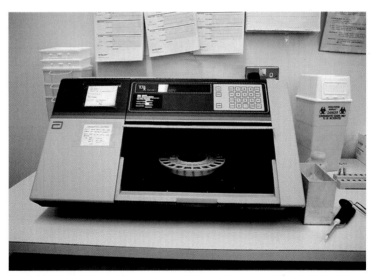

Fig. 34 Automated assay system capable of rapidly measuring concentrations of several different antibiotics.

7 / Urinary tract infection

Urine specimens from patients with cystitis and, to a lesser extent, pyelonephritis account for a large proportion of all specimens sent to the diagnostic laboratory. The laboratory diagnosis of urinary tract infection (UTI) is based on:

- the presence of bacteria (bacteriuria)
- the presence of leucocytes (pyuria).

The laboratory can also guide treatment by:

- identification of bacteria causing infection
- antibiotic susceptibility tests
- confirmation of eradication.

Specimens Many urine specimens do not contribute to the diagnosis or management of UTI because of the presence of contaminating bacteria. Precautions can be taken during specimen collection to reduce the risk of contamination:

- prior cleansing of external genitalia
- avoiding contact with inside of specimen jar
- collection of midstream urine.

Special arrangements may be required for the collection of satisfactory specimens from certain groups of patients:

- *Babies*
  - collection bag (Fig. 35)
  - suprapubic aspiration.
- *Older children*—clean catch specimen.
- *Catheterised patients*—collection via proximal sampling port (Fig. 36).
- *Pyelonephritis*—by ureteric cannulation.

Transport Spoilage of specimens held in transit for lengthy periods can be reduced by:

- borate-containing specimen jar
- refrigeration during transit.

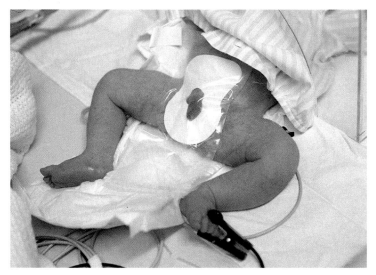

Fig. 35 Paediatric urine collection bag fixed to patient's perineum.

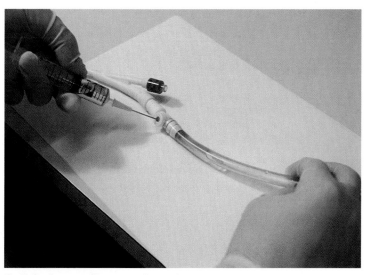

Fig. 36 Specimen sampling port on urinary catheter.

Microscopy   Commonly used microscopic methods:
- wet film
- counting chamber
- inverted microscope (Fig. 37).

Some results are available immediately:
- leucocyte count (Fig. 38)—expressed as cells per ml or cells per high power field
- epithelial cells—suggests contamination
- presence of organisms, if in large numbers.

Pyuria

The presence of leucocytes in urine specimens is often taken to indicate UTI. However, pyuria is not always present during UTI and may be due to non-infective causes.

Pyuria with no bacterial growth may be due to:
- other source of leucocytes (e.g. menstruation)
- prior use of antibiotics
- organisms that do not grow on routine media:
  - growth factor-requiring bacteria
  - anaerobic bacteria
  - mycobacteria.

Bacteriuria but no pyuria can be present in:
- early stages of UTI
- neutropenia
- UTI in elderly patients
- some UTIs due to *Proteus* sp.

Other applications of urine microscopy include:
- casts, e.g. in glomerulonephritis
- crystals.

Fig. 37 Inverted microscope set up to examine urine specimens in microtitre tray.

Fig. 38 Urine specimen with many leucocytes (pyuria).

7 / Urinary tract infection

Culture **Bacteriuria**
A known volume of urine is spread on solid media. Commonly used agars include:
- CLED
- MacConkey
- blood or chocolate agar (optional).

Agar plates are incubated overnight in air at 37°C and then examined for:
- number of colonies on agar (Fig. 39)
  (for 1µl sample, 100 colonies = 10^5/ml)
- purity of growth.

Bacteria commonly isolated include:
- MSU
  - *E. coli*
  - other *Enterobacteriaceae* (coliforms)
  - coagulase negative staphylococci.
- CSU
  - *Enterococcus faecalis*
  - coagulase negative staphylococci
  - *Proteus* spp.
  - *Pseudomonas aeruginosa*.

Reasons for mixed bacterial growth (Fig. 40) include:
- contamination during specimen collection
- catheter or bag specimen
- vesico-colic fistula.

Significant bacterial growth is often regarded as a count greater than 10^5/ml. Important exceptions include:
- Mixed bacterial growth, as above.
- Significant numbers less than 10^5/ml:
  - pyelonephritis.
  - staphylococcal UTI
  - chronic, recurrent UTI
  - 'sterile' collection (e.g. SPA).

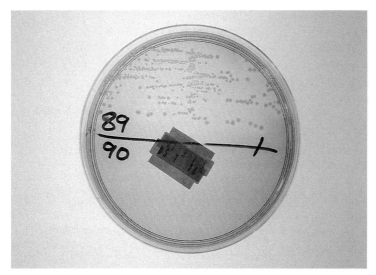

Fig. 39 Heavy growth of 'coliform' on half CLED plate from specimen of urine.

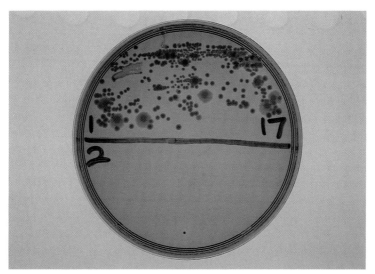

Fig. 40 Mixed bacterial growth on half CLED plate due to contamination with perineal flora.

Supplementary tests

In cases of sterile pyuria, additional investigations may determine the cause:

- *Presence of antibiotics in specimen.* This can be shown by using the urine as a source of antibiotic in a simplified version of the antibiotic bioassay, where any antibiotic present inhibits the growth of a control organism (producing a zone of growth inhibition as in Fig. 41).
- *Early morning urine (EMU).* If renal tuberculosis is suspected, three early morning specimens of urine should be collected to increase the probability of recovering mycobacteria.
- *Fastidious bacteria:* additional culture media may be required. Most laboratories will only attempt to isolate these organisms after discussion with the referring clinician (Fig. 42).

A urine pregnancy test should be performed if pregnancy is suspected. *Blood culture* should always be performed (pyrexial patient).

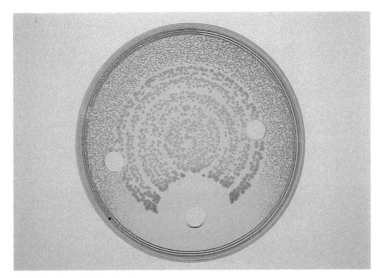

Fig. 41 Test for antibiotics in urine.

Fig. 42 Growth of fastidious bacteria from urine specimen present on chocolate agar, but not on CLED.

7 / Urinary tract infection

Antimicrobial susceptibility

Bacterial resistance to antibiotics used in the treatment of UTI is common. Resistance by a given species to a specific antibiotic agent may be difficult to predict. Susceptibility testing is therefore performed on all significant bacterial isolates.

Common resistance problems include:
- *Enterobacteriaceae* — ampicillin (Fig. 43)
- *Proteus* sp. — nitrofurantoin
- *Pseudomonas* sp. — first-line UTI agents (Fig. 44).

Antibiotic susceptibility results can be obtained by a 'direct' test put up at the same time as preliminary culture, but results may be difficult to read due to mixed growth or heavy inoculum density (Fig. 45). If disc diffusion tests have to be performed with bacteria from the primary culture plate, results usually take a further 24 hours.

Diagnostic laboratories will often recommend a selection of antibiotics, based on:
- activity against bacteria isolated
- suitability for specific type of UTI
- least likely to promote antibiotic resistance.

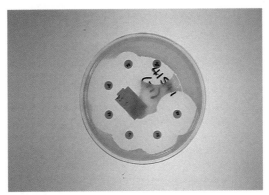

Fig. 43 Ampicillin-resistant coliform isolated from urine specimen. (disc at 1 o'clock).

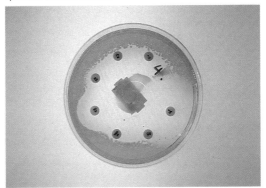

Fig. 44 Characteristic susceptibility pattern of *Pseudomonas* sp. isolated from urine.

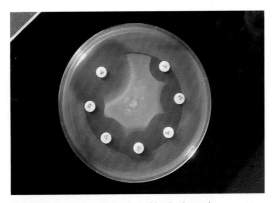

Fig. 45 Direct urine sensitivity plate with mixed growth.

8 / Sexually transmitted diseases

Most diagnostic laboratories restrict their service to the common sexually transmitted diseases: gonorrhoea, chlamydial infection, trichomoniasis, candidiasis and syphilis. Tests may be available through reference laboratories for other sexually transmitted diseases.

Specimens *N. gonorrhoeae* and *Chlamydia trachomatis* are both fastidious organisms that require special care in specimen collection and during transport. The most common specimens submitted for isolation of these two organisms are urethral swabs (men) and cervical swabs. The following procedure should be followed:

Urethral swab
1. Use narrow wire-mounted swab (Fig. 46).
2. Swab urethral discharge, when present.
3. Insert swab into distal urethra.
4. Wait several seconds to absorb exudate.
5. Rotate 360°—dislodges epithelial cells for *Chlamydia*.
6. Prepare smear on microscope slide (Fig. 47).
7. Inoculate agar plate with swab (or place swab in transport medium; Fig. 48).

Cervical swab
1. Obtain swab of endocervix with aid of speculum.
2. Avoid contact with vaginal wall.
3. Prepare microscope slide as above.

Specimens should be obtained from other body sites according to the patient's or contact's history (e.g. rectum, oropharynx and, in women, the urethra).

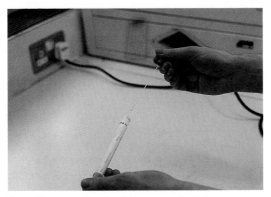

Fig. 46 Wire-mounted swab suitable for collecting inflammatory exudate from the male urethra.

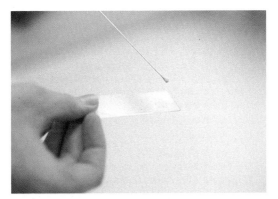

Fig. 47 Smear preparation with urethral pus for immediate microscopy in the clinic.

Fig. 48 Wire-mounted swab cut off into transport medium.

Transport   Liquid transport media are used to assist the survival of STD-causing pathogens during transit from the clinic:
- *N. gonorrhoeae* — Stuart's medium. Inoculation onto selective gonococcal culture agar should take place within 6 hours of sample collection.
- *C. trachomatis* — chlamydia transport medium. This medium should not be refrigerated, and should be inoculated immediately.

Microscopy   **Urethral smear**
Smear preparations of urethral exudate can be used to detect *N. gonorrhoeae* by Gram stain, or *C. trachomatis* by direct immunofluorescence.
- Gram stain
  - neutrophil polymorphs are present in acute gonococcal urethritis, and in non-specific urethritis
  - Gram-negative cocci (especially if intracellular) are suggestive of gonorrhoea (Fig. 49).
- Immunofluorescence — specific test for detection of *C. trachomatis*.

Chlamydial initial bodies fluoresce under ultraviolet light (Fig. 50).

Cervical smear
Gram stain of material sampled from the cervix or vagina rarely contributes to a diagnosis of gonorrhoea.

Chancral exudate
Exudate from the base of a syphilitic chancre can be examined for spirochaetes using dark ground microscopy, on the rare occasions when syphilis presents at this stage.

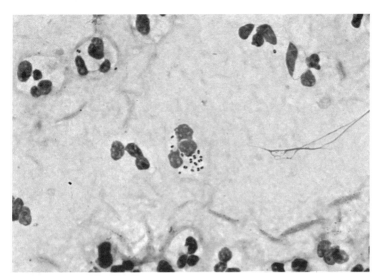

Fig. 49 Gram-stained smear of urethral exudate containing neutrophils and intracellular Gram-negative diplococci strongly suggestive of gonorrhoea.

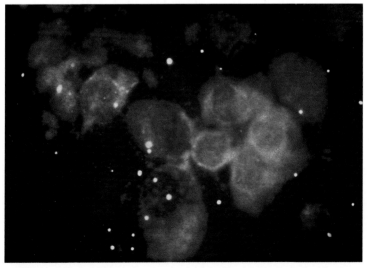

Fig. 50 Fluorescing initial bodies of *C. trachomatis*, labelled with specific immunofluorescent reagent.

Culture Special methods have been developed to isolate the fastidious organisms that cause sexually transmitted diseases from sites that normally have a resident bacterial flora:
- *N. gonorrhoeae* requires an enriched medium (VCAT) and a CO_2 supplemented atmosphere for growth. Antibiotics incorporated in the agar prevent the growth of most commensal bacterial species, but growth may take up to 48 hours (Fig. 51).
- *C. trachomatis* requires cell culture for cultivation because it is an obligate intracellular parasite (Fig. 52).
- *Candida albicans* is a yeast and will grow on Sabouraud's agar. The high glucose concentration and antibiotics prevent the growth of most bacteria (Fig. 53).
- *Trichomonas vaginalis* can be grown in a broth medium.

Fig. 51 *N. gonorrhoeae* growing on VCAT agar.

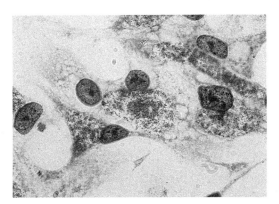

Fig. 52 Perinuclear *C. trachomatis* inclusion bodies in cell monolayer.

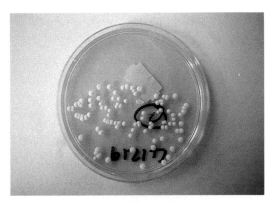

Fig. 53 *C. albicans* colonies on Sabouraud's agar.

Confirmatory tests

N. gonorrhoeae is confirmed using one or more of the following tests:

- conventional carbohydrate tests (CTS sugars)
- rapid carbohydrate tests (Quadferm)
- immunological test (coagglutination).

Carbohydrate utilization tests

Classical carbohydrate utilization tests require 24–48 h from inoculation with suspected *N. gonorrhoeae* (Fig. 54). Carbohydrates used include glucose (G), maltose (M), sucrose (S) and lactose (L). The following results may be obtained:

Organisms	G	M	S	L
N. gonorrhoeae	+	–	–	–
N. meningitidis	+	+	–	–
N. lactamica	+	+	–	+
*Moraxella catarrhalis**	–	–	–	–

* Formerly *Branhamella catarrhalis*

The rapid carbohydrate utilization test (Quadferm) produces a result in several hours and incorporates a β-lactamase test, which gives useful preliminary information on penicillin susceptibility. The results are interpreted as in the conventional test, above (Fig. 55).

Coagglutination test

N. gonorrhoeae can also be confirmed immunologically using a specific coagglutination reagent (Fig. 56). The test takes minutes to perform. When conducted with controls, it is an acceptable alternative to carbohydrate reactions.

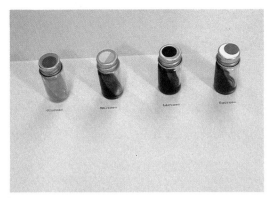

Fig. 54 Conventional sugar reactions for *N. gonorrhoeae*.

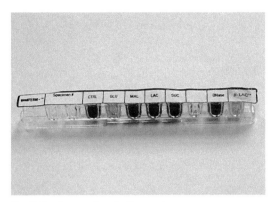

Fig. 55 Quadferm reactions for *N. gonorrhoeae*.

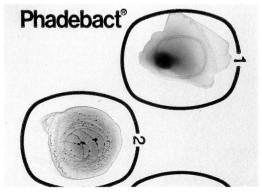

Fig. 56 Coagglutination test for *N. gonorrhoeae*.

Antibiotic susceptibility

Antibiotic resistance is common in *N. gonorrhoeae*. Several types of resistance may be encountered:
- reduced penicillin susceptibility (raised MIC)
- penicillinase production
- tetracycline resistance
- quinolone resistance.

A controlled disc diffusion technique on special sensitivity agar is used to test gonococcal antibiotic susceptibilities. Penicillinase production usually results in complete absence of the penicillin zone (Fig. 57), and can be confirmed by a β-lactamase test. Reduced penicillin susceptibility without penicillinase production may cause partial reduction in the zone diameter, as may the more recently recognized quinolone resistance.

β-lactamase test

Production of penicillinase, a β-lactamase enzyme responsible for destruction of penicillin and consequent loss of its antibiotic property, can be detected in the laboratory using simple tests. Most versions of the β-lactamase test rely on a rapid colour reaction (Fig. 58). Results can be used to give preliminary information on penicillin susceptibility before the results of formal susceptibility testing become available.

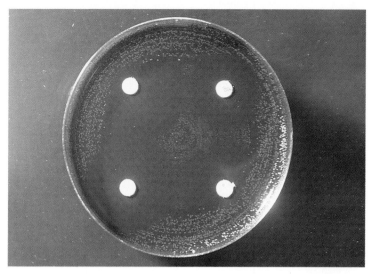

Fig. 57 Complete absence of zone to penicillin with penicillinase-producing *N. gonorrhoeae* (disc at lower right).

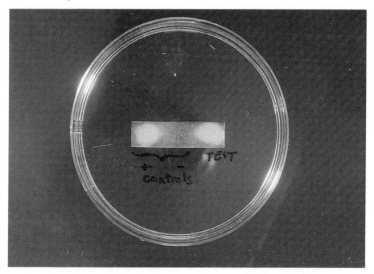

Fig. 58 Positive β-lactamase test (with positive and negative controls) produced by *N. gonorrhoeae* featured in Figure 57 above.

Serological tests

Syphilis

Diagnosis of syphilis relies largely on serological tests. These are:

- VDRL—Venereal Diseases Reference Laboratory test
- TPHA—*Treponema pallidum* haemagglutination assay
- FTAabs—absorbed fluorescent treponemal antibody.

VDRL test. This test relies on non-specific flocculation by antibodies to group antigens which cross-react with mitochondrial lipids (Fig. 59). It becomes positive early in the course of syphilis, rises to high titre during active disease and falls with treatment. Though it is a useful indicator of disease progress, it may be positive in other conditions (biological false positives) such as infectious mononucleosis, lupus erythematosus and leprosy (see table below).

TPHA. This test (Fig. 60) detects agglutinating antibodies to *T. pallidum*-coated erythrocytes. Positive results occur in late primary and secondary syphilis. The TPHA is negative in conditions causing a biological false positive VDRL.

FTAabs. The FTA test employs microscope slides pre-coated with treponemes to bind antibody from the test serum, which will then bind fluorescein-labelled antibody (Fig. 61). The test becomes positive early in primary syphilis and may remain so for many years after successful treatment.

Stage of syphilis	VDRL	TPHA	FTA
Early	+	−	+
Most stages	+	+	+
Late/treated	−	+	+
BFP	+	−	−

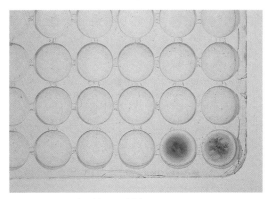

Fig. 59 VDRL test (positive on right).

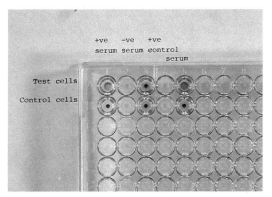

Fig. 60 TPHA.

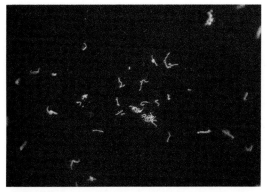

Fig. 61 FTAabs.

9 / **Gastrointestinal tract infections**

Bacterial gastrointestinal infections are usually diagnosed by isolation of the causative agent from a specimen of faeces. The major drawback to this approach is the abundance of commensal organisms in faeces.

Specimens Faecal specimens should be collected into a wide-mouthed, screw-topped container with a small spoon attached to the lid (Fig. 62). Liquid faeces specimens should fill around one-fourth of the container. Solid specimens should be sampled with the spoon: only a small lump is required. Overfilling specimen containers can cause a serious infection hazard. Rectal swabs should not be collected from patients with diarrhoea. Blood cultures should be collected from patients with suspected enteric fever.

The following information should be provided on the specimen request form:
- history of recent travel: when and where
- symptomatic contacts
- whether food handler
- whether pyrexial.

Transport Faecal specimens should reach the laboratory within about 4 hours of collection. If longer, some pathogens may not survive without refrigeration.

Microscopy Microscopy is occasionally used for early diagnosis of *Campylobacter* infection (Fig. 63). Microscopic techniques are used extensively in the diagnosis of enteric parasite infections (q.v.).

Fig. 62 Faecal specimen container.

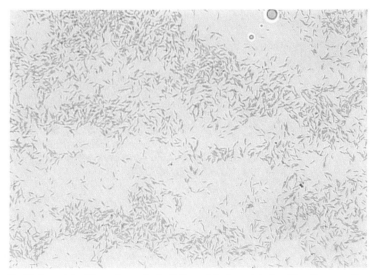

Fig. 63 Gram stain of faeces from patient with *Campylobacter* sp. infection. Many Gram-negative spiral forms are present.

Culture The abundant commensal bacterial flora in faeces causes problems with the recovery of bacterial pathogens. Selective agar media are used to suppress the growth of commensal species while allowing the growth of bacterial pathogens. No single medium is ideal. Most diagnostic laboratories use a combination of at least two of the following selective media:

- XLD—xylose–lysine decarboxylase agar
- DCA—deoxycholate citrate agar (Fig. 64)
- MAC—MacConkey agar.

Preston medium or VCA (vancomycin–colistin–amphotericin) is used to detect *Campylobacter* sp. (Fig. 65). Media used less frequently (when a specific pathogen is suspected) include:

- TCBS (thiosulphate–bile salt–sucrose) for *Vibrio* sp.
- MAC-s (sorbitol–MacConkey) for enterohaemorrhagic *E. coli*
- CIN (cefsulodin–irgasan–novobiocin) for *Yersinia* sp.

Enrichment broths are used to selectively encourage the growth of very low numbers of faecal pathogens. Those in common use include:

- Selenite broth for *Salmonella* sp.
- alkaline peptone water for *Vibrio* sp.

Agar plates and enrichment broth are incubated at 37°C in air, except for *Campylobacter* sp. which requires 10% CO_2 and reduced O_2 content at 43°C. Preliminary screening of suspect colonies can normally be performed after overnight incubation. *Campylobacter* sp. may require a further 24 h incubation.

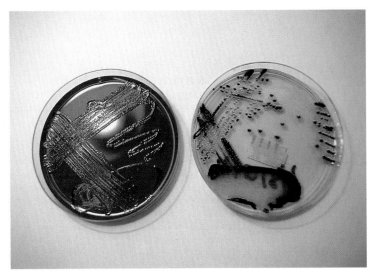

Fig. 64 *Salmonella* sp. grown on XLD and DCA agars.

Fig. 65 *Campylobacter* sp. growing on VCA agar.

Preliminary identification

Some bacteria not associated with gastrointestinal infections resemble pathogenic species on selective media (e.g. *Proteus* sp. which may resemble *Salmonella* sp.). A urease test is performed to exclude *Proteus* sp. (Fig. 66). The suspected pathogen can then be tested with polyvalent agglutinating antisera in a slide agglutination test (Fig. 67). Suspected *Campylobacter* sp. colonies are tested using the oxidase test and Gram stain.

A provisional report is often produced as soon as there is a degree of confidence that a bacterial pathogen has been isolated. These reports may be substantially modified by the results of confirmatory tests. Provisional reports are based on the following:

- *Salmonella* sp.
  - non-lactose fermenting colonies
  - often H_2S positive (black on XLD)
  - urease negative
  - *Salmonella* poly-O positive.
- *Shigella* sp.
  - non-lactose fermenting colonies
  - urease negative
  - *Shigella* sp. O antigen positive
- *Campylobacter* sp.
  - growth at 43°C
  - Gram-negative/spirals
  - oxidase positive.

Confirmatory tests usually require at least a further 24 h before results are known.

Fig. 66 Positive urease broth after inoculation with *Proteus* sp. and negative control.

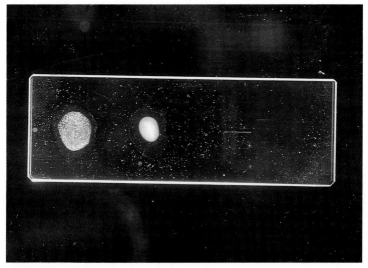

Fig. 67 Positive reaction with polyvalent O antigen salmonella agglutinating antiserum and negative control.

Confirmatory tests

The definitive identification of *Salmonella* and *Shigella* species requires both biochemical and serological tests.

Biochemical identification of bacteria

Single colonies of provisionally identified bacterial pathogens are used to perform a combination of utilization and fermentation tests. Tests may be performed in separate containers as a 'short set', or with a multiple test system (Fig. 68).

Identification of serotype

Agglutinating antisera are used to identify the serotype of bacterial gastrointestinal pathogens. Slide agglutination results may need confirmation by a tube agglutination test in which a suspension of the organisms is treated to extract or immobilize the antigen and then incubated with serial dilutions of the relevant antisera (Fig. 69). Most diagnostic laboratories carry a limited stock of the most commonly used antisera. More detailed identification of serotypes requires a large range of antisera and should be performed in a reference laboratory.

Antisera agglutinate:

- common *Salmonella* O antigens
- common *Salmonella* H antigens
- Vi antigen (*S. typhi*)
- *Shigella* O antigens
- common enteropathogenic *E. coli* O antigens.

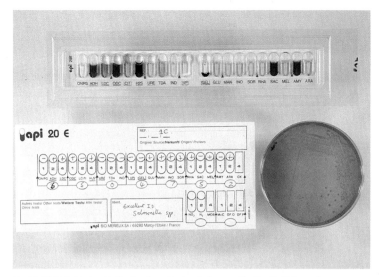

Fig. 68 Multiple biochemical test gallery after incubation used to identify *Salmonella* sp.

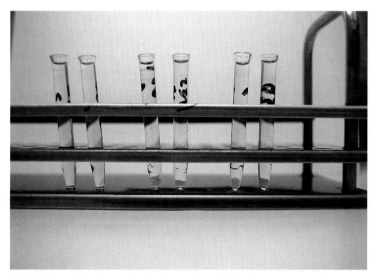

Fig. 69 Tube agglutination reaction for selected salmonella H antigens.

9 / Gastrointestinal tract infections

Plasmid typing
During a suspected outbreak of bacterial gastrointestinal infection caused by a common serotype (e.g. *S. enteritidis*) it is necessary to obtain further evidence that bacteria from different patients are indistinguishable. Plasmid typing by polyacrylamide gel electrophoresis can be used for this purpose (Fig. 70).

Toxin production
Tests for bacterial toxin production are, with few exceptions, not widely used. Commoner tests systems include:
- cytopathic effect on a cell culture monolayer (Fig. 71)
- antigen detection (e.g. latex agglutination, ELISA)
- DNA probes.

Toxins more commonly sought include:
- *Clostridium difficile* toxin in pseudomembranous colitis
- *E. coli* enterotoxin in traveller's diarrhoea.

Antimicrobial susceptibility

Antimicrobial susceptibility tests are not usually performed on bacteria isolated from faecal specimens since antibiotics are not normally required for uncomplicated gastroenteritis. Susceptibility tests are performed in the following cases:
- to assist identification of *Campylobacter* spp. (Fig. 72)
- enteric fever: antibiotic treatment is required and resistance is common
- patients at extremes of age with suspected bacteraemia.

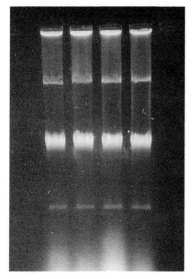

Fig. 70 Plasmid gel electrophoresis showing indistinguishable banding patterns for salmonellas isolated from four different patients.

Fig. 71 Cytopathic effect of *C. difficile* toxin on cell monolayer.

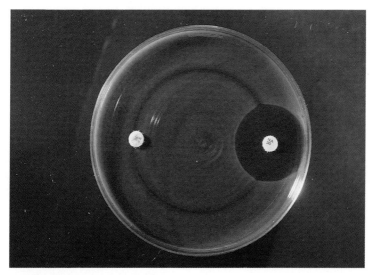

Fig. 72 *Campylobacter* sp.: nalidixic acid and cephaloridine susceptibility tests to assist further identification.

10 / **Respiratory tract infections**

The abundant microbial flora of the upper respiratory tract often causes difficulty in distinguishing bacterial pathogens and commensal flora. The contribution of the diagnostic laboratory may be restricted by the poor quality of specimens.

Specimens

Throat swab
Many cases of acute pharyngitis (including cases of uncomplicated streptococcal pharyngitis) probably do not require antibiotic treatment. Throat swabs are more helpful in cases of tonsillitis or peritonsillar abscess, but must be taken carefully, avoiding contact with the tongue (Fig. 73).

Sputum specimen
Sputum should be collected into a sterile, wide-mouthed container with a secure lid. The specimen should be coughed, which may require the aid of a physiotherapist; specimens composed of saliva are of little diagnostic value. The macroscopic appearance of specimens is often described as follows:
- *salivary*—clear/runny (Fig. 74)
- *mucopurulent*—intermediate (may have flecks of pus)
- *purulent*—green or yellow/thick (Fig. 74)
- *bloodstained*.

Bronchial lavage specimens (which resemble salivary specimens) must be clearly labelled.

Blood cultures
These should always be taken from patients with fever and a suspected lower respiratory tract infection.

Fig. 73 Tongue depressor used to prevent contact with tongue or buccal mucosa while collecting a throat swab.

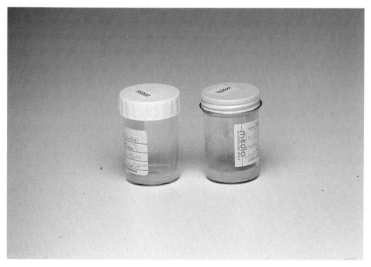

Fig. 74 'Sputum' specimen composed of saliva (left). Purulent sputum specimen (right).

Microscopy The abundant bacterial flora of the upper respiratory tract makes microscopy unhelpful, except in the examination of pus (e.g. peritonsillar abscess).

Gram stains of sputum are routinely examined for their cellular and bacterial content. Neutrophil polymorphs, buccal epithelial cells or alveolar cells may be seen (Fig. 75). Large numbers of neutrophils contribute to the purulent appearance, while abundant epithelial cells indicate salivary contamination. Some laboratories use the polymorph:epithelial cell ratio to assess the suitability of sputum specimens for culture.

The recognition of a single type of organism in the Gram stain can occasionally help an early decision on the choice of antibiotic agent (see table below).

Gram-stain appearance	Possible pathogen
Gram +ve diplococci	*S. pneumoniae* (Fig. 76)
Small Gram −ve coccobacilli	*H. influenzae* (Fig. 77)
Gram +ve cocci (clusters)	*S. aureus*

Stains for acid-fast bacilli
Zeihl–Neelsen or auramine stains (q.v.) are used to examine sputum smears for acid-fast bacilli (Fig. 78).
- A positive smear does not:
  - confirm a diagnosis of tuberculosis
  - identify species of mycobacteria
  - always signify viable organisms.
- A negative smear does not rule out a diagnosis of pulmonary tuberculosis.

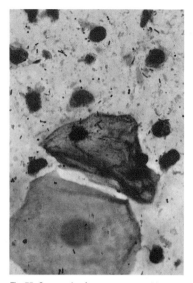

Fig. 75 Gram stain of sputum smear with many neutrophil polymorphs and some epithelial cells.

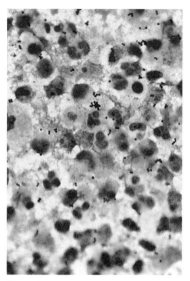

Fig. 76 Gram-positive diplococci in sputum smear, suggestive of *S. pneumoniae*.

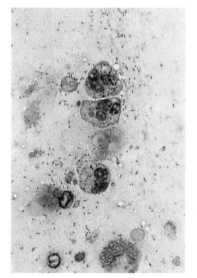

Fig. 77 Small Gram-negative coccobacilli in sputum smear, suggestive of *H. influenzae*.

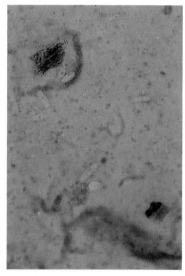

Fig. 78 Acid-fast bacilli in Zeihl–Neelsen stained sputum smear.

Culture **Throat swab**
Oropharyngeal swabs are cultured on blood agar.
A tellurite agar plate is also used when diphtheria is suspected. The major bacterial pathogens sought include:
- *Streptococcus pyogenes* (Lancefield group A)
- *Corynebacterium diphtheriae*
- *Corynebacterium ulcerans*.

S. pyogenes will be considered whenever beta haemolytic colonies are present on the blood agar plate (Fig. 79). *Corynebacterium diphtheriae* has a typical appearance on tellurite medium which may allow presumptive identification to biotype level (Fig. 80).

Sputum
Purulent sputum is inoculated onto blood and chocolate agar plates and incubated at 37°C in a moist atmosphere containing 5% CO_2. A streak of *S. aureus* culture and an optochin disc may be added to the blood agar plate to assist rapid identification of *S. pneumoniae* (Fig. 81) and *H. influenzae* (Fig. 82). Sputum from hospitalized patients at risk from nosocomial pneumonia may also be cultured on MacConkey agar to help speed identification of Gram-negative species. Preliminary results may be available after overnight incubation, but definitive identification may require at least another 24 hours.

If culture for mycobacteria has been requested, sputum will also be treated to reduce contamination by other bacteria, and inoculated onto Lowenstein–Jensen slopes and into Kirschner's broth.

Fig. 79 Bacterial growth from throat swab on blood agar with beta haemolytic colonies.

Fig. 80 Colonies of *Corynebacterium diphtheriae* on tellurite medium.

Fig. 81 Colonies of *S. pneumoniae* on primary culture plate, inhibited by optichin.

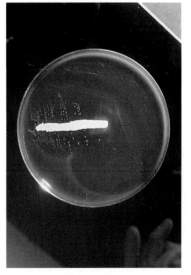

Fig. 82 Preferential growth of *H. influenzae* colonies around *S. aureus* streak.

Confirmatory tests

Streptococcal grouping
A latex agglutination technique (Fig. 29, p. 20) is used to determine the group specific antigen (A, B, C, D, F and G) present in an enzyme extract of beta haemolytic streptococci.

Bile solubility
S. pneumoniae is lysed by bile salt solution after exposure at 37°C.

Optichin sensitivity
A zone of inhibition around an optichin disc indicates the presence of *S. pneumoniae*. The test may be incorporated in the primary isolation plate (Fig. 81, p. 62).

Slide coagulase
The presence of *S. aureus* is rapidly confirmed by clumping in the presence of rabbit plasma (Fig. 26, p. 18).

Haemophilus growth factor tests
Haemophilus spp. usually require growth factors, X factor (haemin) and/or V factor (NAD), the requirement for which is routinely tested by disc diffusion. *H. influenzae* requires both X and V factors (Fig. 83), while *H. parainfluenzae* requires only V factor (Fig. 84).

Biochemical activity
Gram-negative bacilli, including *P. aeruginosa*, if considered to be the cause of nosocomial pneumonia, may require further identification using a panel of biochemical tests such as API20NE (non-enteric) (Fig. 85).

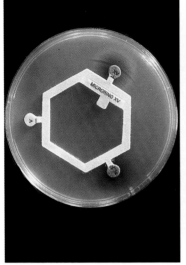 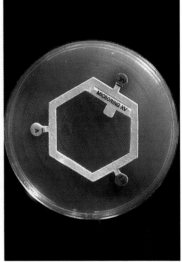

Fig. 83 Growth factor requirement test in which *H. influenzae* has only grown around the disc containing X and V factors.

Fig. 84 *H. parainfluenzae* has grown around both the XV disc and the V disc.

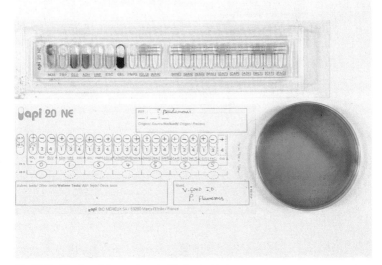

Fig. 85 Identification of *P. aeruginosa* by biochemical tests.

10 / Respiratory tract infections

Antibiotic susceptibility

β-lactamase is important as a mechanism of ampicillin resistance in a proportion of *H. influenzae* clinical isolates. Its presence can be demonstrated rapidly using a colorimetric strip test (see p. 43) before antibiotic sensitivity testing has been carried out, or to confirm sensitivity results (Fig. 86).

Resistance to commonly used antibiotics is often found in bacteria causing respiratory infections. Antibiotic sensitivity testing is a necessary sequel to laboratory confirmation of bacterial chest infections. Commoner resistance problems include:

- ampicillin-resistant *H. influenzae*
- erythromycin-resistant *S. pneumoniae* (Fig. 87)
- penicillin-resistant *S. aureus*
- multiply-resistant *P. aeruginosa*.

Most diagnostic laboratories will test the following agents or alternatives:

- benzyl penicillin
- ampicillin
- erythromycin
- tetracycline.

Diagnostic laboratories will differ in their choice of other antimicrobial agents tested, particularly those used to treat infections caused by resistant bacteria or Gram-negative species.

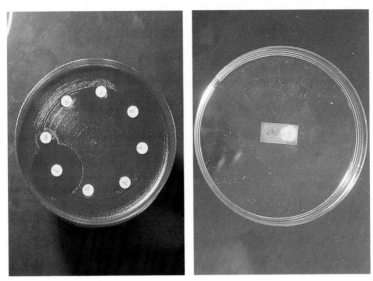

Fig. 86 Ampicillin resistant *H. influenzae* and corresponding β-lactamase test.

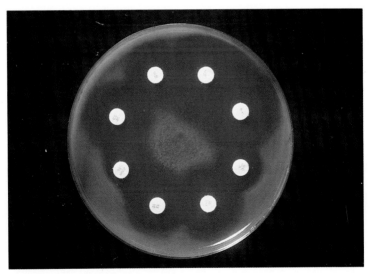

Fig. 87 Erythromycin-resistant *S. pneumoniae*.

11 / Meningitis

Members of all major classes of microorganisms can cause intracranial infection. The microbiological diagnosis of cerebral abscess is similar to that of soft tissue infections (q.v.). Encephalitis requires biopsy or diagnostic tests available in reference laboratories. This section is therefore restricted to meningitis.

Specimens   Cerebrospinal fluid (CSF) should be collected by lumbar puncture with careful skin disinfection and using a sterile technique. CSF should be collected into three sterile, numbered, plastic containers for bacteriology, and further containers for protein and glucose estimations.

Blood culture should be performed at the same time. A further blood sample should be collected for blood glucose, to compare with CSF level.

Main causes of blood-stained CSF include:

- blood contamination during collection (Fig. 88) (RBC count may fall in bottles 2 and 3)
- subarachnoid haemorrhage (Fig. 89) (RBC count stays high in bottles 1–3).

CSF turbidity is not always due to leucocytes or bacteria.

Transport   All CSF specimens should be transported to the laboratory quickly to avoid an artifactual fall in leucocyte count, or death of fastidious pathogens. In some centres, glucose and protein estimations need to be sent to another laboratory (e.g. clinical chemistry).

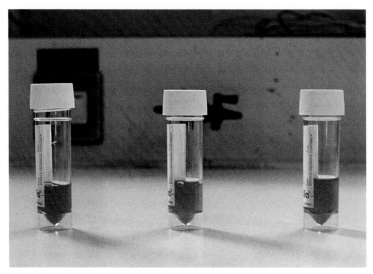

Fig. 88 Blood-stained CSF due to traumatic lumbar puncture.

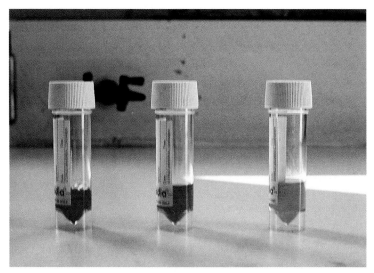

Fig. 89 Blood-stained CSF due to subarachnoid haemorrhage.

Microscopy CSF is normally sterile and almost cell free. Any visible bacteria or significant rise in cellular content is significant.

Leucocytes in CSF
Total leucocyte count in uncentrifuged CSF is obtained by counting cells manually in a haemocytometer chamber under the light microscope (Fig. 90). A differential count is obtained by staining a centrifuged CSF deposit (Fig. 91). Differential counts are not helpful if:

- very small amount of CSF for examination
- very few leucocytes present
- clotted specimen.

Bacteria in CSF Gram stain
The Gram stain is used to detect bacteria in fresh, centrifuged CSF. It requires around 10^5–10^6 bacteria per ml. Acridine orange stain may be more sensitive.

Gram stain	Possible pathogen
Positive diplococci	*S. pneumoniae*
Negative coccobacilli	*H. influenzae*
Negative diplococci	*N. meningitidis* (Fig. 92)
Positive cocci, chains	Streptococci (e.g. group B)
Positive bacilli	*Listeria monocytogenes*
Negative bacilli	'Coliforms' (e.g. *E. coli*)

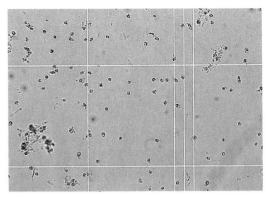

Fig. 90 Leucocytes in CSF from patient with meningitis, counted in haemocytometer.

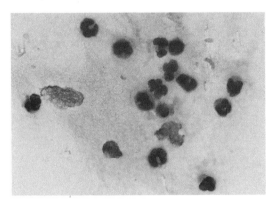

Fig. 91 Stained centrifuge deposit of CSF for differential leucocyte count. Neutrophils and lymphocytes are present.

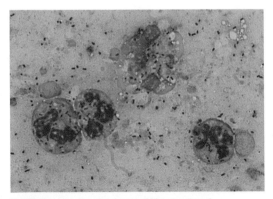

Fig. 92 Gram-negative diplococci in CSF suggestive of *N. meningitidis*.

11 / Meningitis

Initial CSF results:

Type	Neutrophils	Lymphocytes	Protein	Glucose	Gram stain
Viral meningitis		50–500/mm^3	0.5–1 g/l	Normal	Nil
Bacterial meningitis	500–2000/mm^3		1–3 g/l	Low	Bacteria may be seen
Tuberculous meningitis	100–600/mm^3		1–6 g/l	Low	Nil

Rapid diagnosis The high mortality and morbidity of bacterial meningitis necessitate rapid diagnosis. Early confirmation of infection is possible by checking:
- CSF lactate level
- CSF endotoxin (Limulus lysate)—positive in Gram negative infection

  An early indication of causal agent can be given with:
- CSF counterimmune electropheresis (CIE): time consuming and not very sensitive (Fig. 93)
- coagglutination: available for most common bacterial agents of meningitis (Fig. 94).

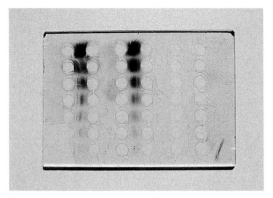

Fig. 93 Counterimmune electrophoresis in which an electric current accelerates precipitation of antigen (in CSF) with antibody from corresponding wells.

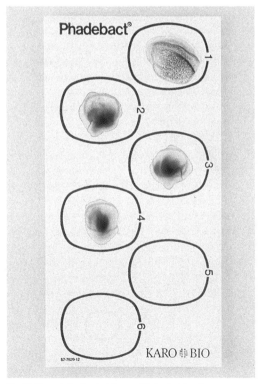

Fig. 94 Coagglutination test for specific bacterial pathogen (positive result top right).

Culture   CSF is inoculated onto non-selective media (e.g. blood agar and chocolate agar) and incubated overnight at 37°C in 5% CO_2. MacConkey agar may also be used if there is any reason to expect meningitis due to *Enterobacteriaceae*.

Any growth must be regarded as significant until proven otherwise. Some bacteria may require prolonged incubation before visible growth appears.

Common bacterial causes

H. influenzae: small Gram-negative coccobacillus that requires X and V factors to grow (growth on chocolate agar but not on blood agar).

S. pneumoniae: Gram-positive diplococcus that is optichin sensitive and bile soluble.

Neisseria meningitidis: Gram-negative diplococcus, (Fig. 95) confirmed by carbohydrate utilization tests (glucose, maltose positive).

Culture-negative meningitis: no bacterial growth on routine culture. Possible causes include:

- Antibiotics prior to lumbar puncture (e.g. by GP). May be coagglutination positive (Fig. 96).
- Cerebral abscess.
- Fastidious or non-cultivatable organisms:
  - viruses
  - *Borrelia burgdorferi* (Lyme disease)
  - mycobacteria
  - *Naegleria* and other amoebic spp.

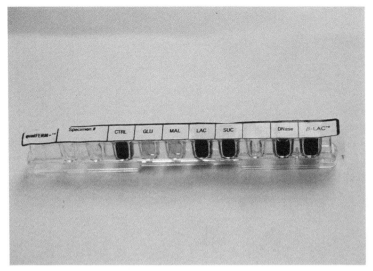

Fig. 95 Carbohydrate reactions for *N. meningitidis*.

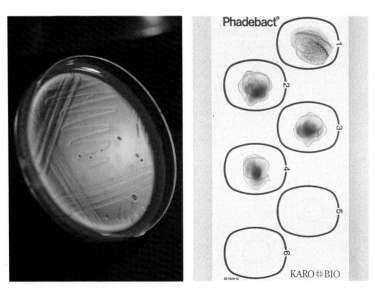

Fig. 96 No growth but positive coagglutination test in partially treated meningococcal meningitis.

11 / Meningitis

Culture (cont)

Less common bacterial causes

The less common causes of meningitis may be suggested by the clinical setting. This may be used to select additional media to help isolate the causative agent. Important clinical associations include:

- newborns—*E. coli*, group B streptococci, *Listeria*
- immunocompromised—*Listeria*
- patients of Asian origin—*M. tuberculosis*
- hydrocephalus (shunt)—coagulase-negative staphylococci, *Enterococcus*.

E. coli will grow on non-selective media used for CSF culture, but may be recognized more rapidly if MacConkey agar is used. Group B streptococci are less haemolytic than group A streptococci, and are easily confirmed by antigen extract latex agglutination. *Listeria monocytogenes* (Fig. 97) may be mistaken for contaminating coryneform bacteria, or for group B (Fig. 98) streptococci. This can be avoided by careful Gram stain and catalase test. Mycobacteria are rarely cultured from patients with tuberculous meningitis, but CSF microscopy (e.g. auramine) and culture for mycobacteria should still be carried out if there is good reason to suspect the diagnosis.

The significance of bacteria grown from shunt CSF specimens may be difficult to assess (Fig. 99), especially if the specimen was taken from a subcutaneous reservoir. Further information about the patient, and repeat specimens may be required.

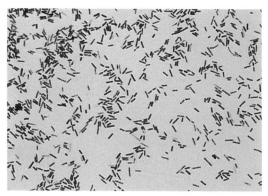

Fig. 97 Gram stain of *Listeria monocytogenes*, which superficially resembles corynebacteria.

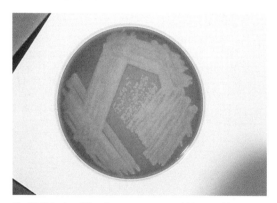

Fig. 98 Colonies of *Listeria monocytogenes* which resemble group B streptococci.

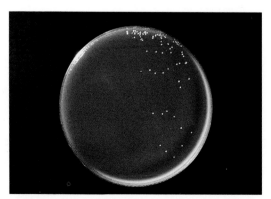

Fig. 99 Colonies of coagulase-negative staphylococci from an intracranial shunt, of indeterminate significance.

11 / Meningitis

Antibiotic susceptibility

Decisions often have to be taken on the most suitable antibiotic before results of CSF culture are known. Empiric therapy may be modified when the causative agent and its antibiotic sensitivity become known.

Antibiotics commonly used in bacterial meningitis include:
- ampicillin
- chloramphenicol
- benzyl penicillin
- cefotaxime.

The commoner antibiotic resistance patterns involve:
- *H. influenzae:* ampicillin, chloramphenicol (Fig. 100)
- *N. meningitidis:* sulphonamides (Fig. 101).

Antibiotic resistance is detected (as with bacterial isolates from other sites) by disc diffusion testing. Results on CSF isolates normally require a further 24 hours after primary isolation of the causative agent. β-lactamase production by ampicillin-resistant *H. influenzae* can be confirmed with the rapid β-lactamase test (Fig. 86, p. 66).

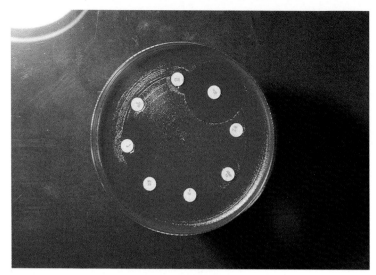

Fig. 100 Ampicillin-resistant *H. influenzae* (disc at 4 o'clock).

Fig. 101 Sulphonamide-resistant *N. meningitidis* (disc at 8 o'clock).

12 / Soft tissue infections

A wide range of sites and causative organisms may be involved in soft tissue infections. At surface sites it may be impossible to distinguish between colonization and true infection.

Specimens Inflammatory exudate (pus) is the specimen most commonly submitted for laboratory examination. Pus is usually collected and transported to the laboratory on a cotton-tipped swab (Fig. 102). If several millilitres of pus or other fluid are available, fastidious organisms (e.g. anaerobic bacteria) survive better if the specimen is sent in bulk in a sterile, screw-topped container (Fig. 103).

When systemic infection is suspected, blood cultures should be collected at the same time.

Sometimes tissue specimens are supplied for microbiological investigation. As large as possible a specimen should be sent. No formalin should come into contact with the specimen.

Transport Swabs, pus and tissue specimens should reach the laboratory as soon as possible to avoid deterioration of the specimen and assist recovery of fastidious bacteria. Material collected during operative procedures requires particular care to ensure that it is not forgotten at the end of the operation (Fig. 104).

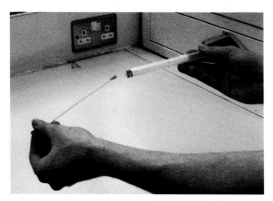

Fig. 102 Cotton-tipped swab with pus sample on tip.

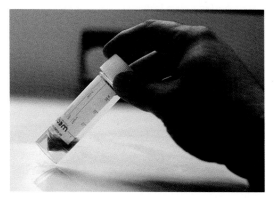

Fig. 103 Screw-topped container with pus sample.

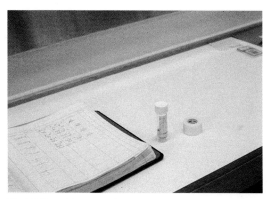

Fig. 104 Bacteriological specimen left behind in operating theatre at end of list.

12 / Soft tissue infections

Microscopy Examination of Gram-stained inflammatory exudate may provide clinically helpful information: recognition of a single type of organism at microscopy (Fig. 105) and may assist decisions on initial antibiotic therapy. Urgent Gram stain may occasionally be necessary to confirm a clinical diagnosis of gas gangrene (clostridial myonecrosis) (Fig. 106).

In some specimens, bacteria demonstrated at microscopy may subsequently fail to grow, due to antibiotic treatment prior to specimen collection.

Rapid diagnosis Antigen detection and other rapid diagnostic techniques are usually unsuited to specimens of inflammatory exudate because of the wide variety of potential bacterial pathogens and large quantities of proteinaceous material.

Some laboratories use gas liquid chromatography (GLC) to detect the presence of anaerobic bacteria, both viable and non-viable, in pus specimens. The GLC detects volatile fatty acids produced by anaerobic bacteria (Fig. 107), but substantial quantities of pus are required.

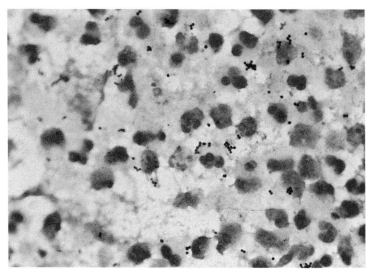

Fig. 105 Gram stain of pus containing Gram-positive cocci, suggestive of *S. aureus*.

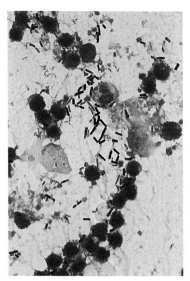

Fig. 106 Gram stain of exudate from patient with gas gangrene, showing Gram-positive bacilli, consistent with *Clostridium* sp.

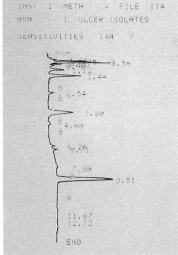

Fig. 107 Gas liquid chromatograph trace showing peaks corresponding to the major volatile fatty acids associated with anaerobic bacteria.

12 / Soft tissue infections

Culture A variety of non-selective and selective media are used to cultivate organisms from soft tissue infections. A common selection would be blood, chocolate, MacConkey and Sabouraud's agars incubated aerobically, and blood agar incubated anaerobically, at 37°C. Unrepeatable specimens (e.g. tissue specimens) may be incubated in Robertson's cooked meat broth to assist the recovery of particularly fastidious organisms.

Antibiotic discs are sometimes placed on primary isolation plates to assist the recognition of important groups of bacteria (e.g. anaerobes; Fig. 108), but rarely help guide antibiotic therapy. The frequency with which more than one bacterial species are isolated from soft tissue infections (Fig. 109) prevents the 'direct' sensitivity test method from being used.

Most clinically important aerobic bacterial species will produce visible growth after overnight incubation. Macroscopic appearance of individual colonies, Gram stain and bench top tests should give an early indication of the organism's identity.

Recognition of bacteria after overnight incubation:

- staphylococci (*S. aureus** and coagulase-negative staphylococci*)
- streptococci (β-haemolytic*, viridans group, non-haemolytic)
- 'coliforms' (*E. coli*, *Klebsiella* sp., *Proteus* sp., etc.)
- pseudomonads (e.g. *P. aeruginosa*)
- fast-growing anaerobes (some *Bacteroides* sp., *Clostridium* sp.).

* May be fully identified at an early stage.

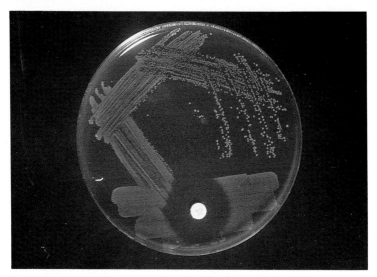

Fig. 108 Anaerobic bacteria present on primary isolation plate, revealed by inhibition zone around metronidazole disc.

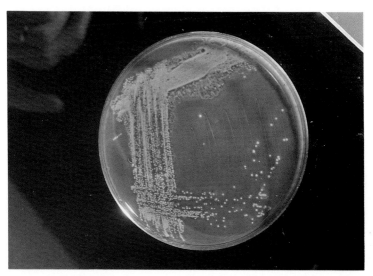

Fig. 109 Mixed bacterial growth (*S. aureus* and *Proteus* sp.) present on primary isolation plate inoculated with wound swab.

Culture (cont)   Some organisms implicated in soft tissue infections require extended incubation. These include:
- Anaerobes
  - *Bacteroides fragilis, Bacteroides melaninogenicus* and other *Bacteroides* sp.
  - some *Clostridium* sp.
- *Pasteurella* sp.
- Microaerophilic streptococci
- Higher bacteria
  - *Actinomyces* sp. (Fig. 110)
  - *Nocardia* sp.

Since definitive identification of most of these organisms may take much longer, a provisional report may be issued at this stage.

Many laboratories do not have facilities to take identification further than a few simple confirmatory tests, such as:

- *Bacteroides spp.* Antibiotic sensitivities are used to distinguish between different members of this group (Fig. 111).
- *Clostridium perfringens*. A Nagler plate is used to detect production of lecithinase, indicated by an opaque zone surrounding the colonies, which is inhibited by specific antitoxin (Fig. 112). Several other clostridia may give a positive result with this test.

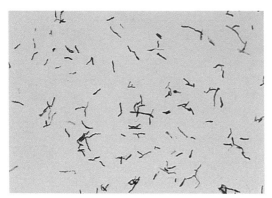

Fig. 110 Gram stain of *Actinomyces* sp.

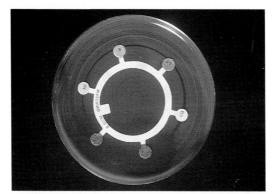

Fig. 111 Preliminary identification of commoner anaerobic species by antibiotic susceptibility.

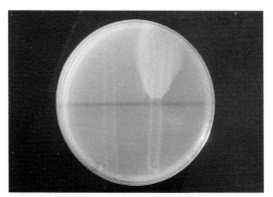

Fig. 112 Nagler plate for demonstration of lecithinase production by some clostridial spp., showing negative and positive controls.

12 / Soft tissue infections

Antibiotic susceptibility

Antibiotic resitance is common to agents used in the treatment of soft tissue infections. Initial antibiotic choice may have to be modified:

- according to results of microscopy
- when causative agent is identified
- when antibiotic sensitivities are confirmed.

Since several bacterial species may be isolated from the same specimen, it is often necessary to first obtain the suspected agent(s) of infection in pure growth by picking single colonies from the primary culture plate (Fig. 113), and using these to inoculate a subculture plate (Fig. 114).

Laboratories vary considerably in the antibiotics they choose to test and report. A wide variety may be tested from which a narrower choice of clinically suitable agents can be made (Fig. 115). In some cases results of antibiotic susceptibility in the laboratory may not be reported because they do not equate with clinical efficacy (e.g. due to poor bioavailability).

Whenever it is not possible to use the antibiotics recommended for use in a given patient, a more satisfactory choice can usually be made as a result of consulting a medical microbiologist.

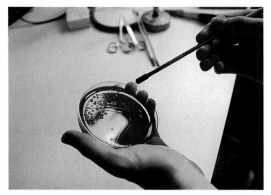

Fig. 113 Single colony pick-off from mixed bacterial growth, using straight wire.

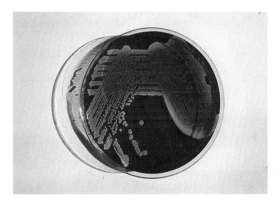

Fig. 114 Pure growth obtained by technique in Figure 113.

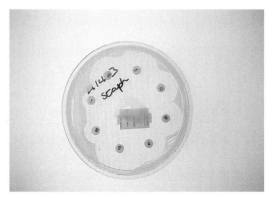

Fig. 115 Full range of antibiotics tested, from which clinically appropriate agents are chosen.

13 / Septicaemia

Bacteria are recovered from the bloodstream of septicaemic patients by culturing venous blood (referred to as 'blood culture'). The high mortality attributed to untreated septicaemia makes the speed of isolation, identification and antibiotic susceptibility testing a high priority.

Specimens The sample is inoculated into bottles of liquid culture medium. Two bottles are used in most centres: one for recovery of aerobic bacteria, and one for anaerobic bacteria (Fig. 116). A high standard of sample collection is essential, using the following approach:

1. skin disinfection at venesection site (Fig. 117)
2. no touch, aseptic venesection technique
3. inoculate bottles after replacing syringe needles
4. inoculate each bottle with optimal blood volume
5. incubate at 37°C immediately.

Other specimens to be considered include:

- Blood cultures
  - at intervals (suspected endocarditis; series of 3)
  - via intravascular cannulas (suspected line sepsis).
- Focus of infection
  - pneumonia (sputum)
  - UTI (MSU/CSU)
  - wound infection (swab/pus)
  - meningitis (CSF)
  - abscess (drained pus)
  - device-related infection (device or contents).

Fig. 116 Set of blood culture bottles: aerobic (left) and anaerobic (right).

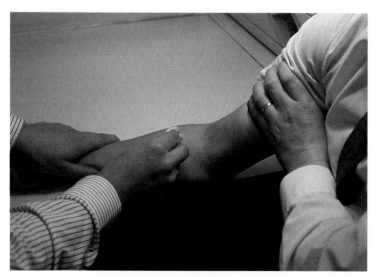

Fig. 117 Skin disinfection prior to venesection for blood culture.

13 / Septicaemia

Detection systems

Manual and automated systems have been devised to encourage rapid microbial growth and speed detection of blood cultures. Growth of positive cultures must be recognized early, and false positives caused by contamination must be avoided. Whatever system is used, accurate screening of large numbers of cultures, many of which will be sterile, is essential.

Manual screening
- Haemolysis.
- Turbidity (Fig. 118).
- Microscopy (Gram stain, acridine orange).
- Routine subculture.

Aids to manual screening
- Headspace device (registers gas pressure increase).
- Agar surface (slope in bottle or separate chamber) (Fig. 119).

Automated screening
Used in the detection of:
- radioactive CO_2
- infrared spectroscopy (Fig. 120)
- turbidity
- conductance change.

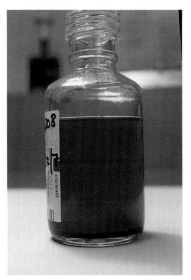

Fig. 118 Turbid blood culture bottle after incubation due to bacterial growth.

Fig. 119 Enclosed agar surface sampling device, attached to blood culture bottle in laboratory.

Fig. 120 Automated blood culture sampling device (Bactec).

Microscopy Detection of microbial growth, particularly in manually screened systems, depends on microscopy. Many clues to the identity of microbial isolates, such as colonial size, shape and colour, are obtained only after growth on solid media and are not obtained when bacteria have been grown in liquid media.

Microscopy provides some initial information as to the identity of blood culture isolates. The microscopic appearance of some bacteria varies between growth in liquid and on solid media (Fig. 121). Many centres restrict their first blood culture reports to a simple description of microscopic appearance, such as:

- Gram-positive cocci
- Gram-negative bacilli.

Gram stain may only detect the presence of bacteria in liquid culture at concentrations greater than or equal to 10^6 per ml. Acridine orange stain is used by some laboratories to improve the sensitivity of detection (Fig. 122), but where this is done, Gram stain has to also be performed in addition to determine the Gram classification.

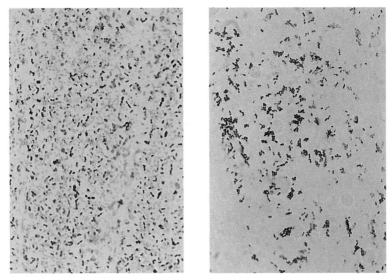

Fig. 121 Gram stain of *S. pneumoniae* grown on agar plate (left), and in blood culture bottle (right).

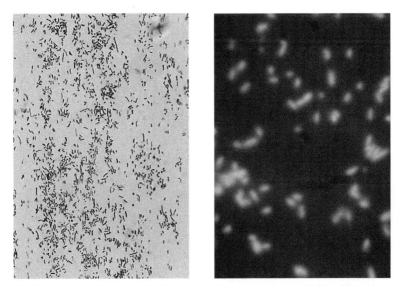

Fig. 122 Gram stain of blood culture bottle contents (left), and the same, stained with acridine orange (right).

13 / Septicaemia

Culture Bacteria from blood culture bottles must be grown on solid, agar-based media before they can be definitively identified. Non-selective media (such as blood agar and chocolate agar) are used for incubation under aerobic or CO_2 enriched conditions, and at least one type of medium is incubated anaerobically.

Bacterial growth after overnight incubation is usually sufficient for confirmatory tests which include the following:

- bench-top tests (e.g. catalase, coagulase, oxidase)
- biochemical tests (e.g. multiple test gallery)
- immunological tests (e.g. meningococcal coagglutination).

Assessment of significance

Commensal skin organisms contaminate a proportion of blood cultures, but under certain conditions these same organisms may cause infections associated with bacteraemia (Fig. 123). Where there is any doubt about the significance of a particular blood culture isolate, the following questions may help:

- Is there a specific association between the organism and the disease?
- Does the isolate account for the patient's disease?
- Is the isolate a common component of skin flora?
- Is this a possible laboratory contaminant?
- Is the isolate present in more than one culture set?
- Has the same organism been isolated from suspected focus of infection (Fig. 124)?

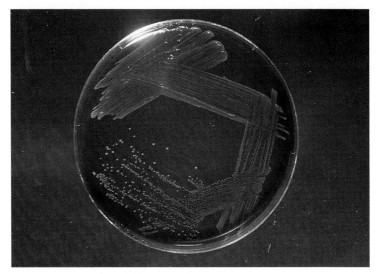

Fig. 123 JK diphtheroid from Hickman line infection in leukaemic, on agar. It is a commensal organism although significant in these circumstances.

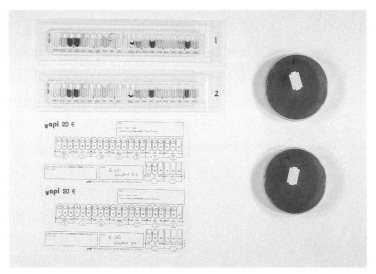

Fig. 124 Bacteraemia associated with UTI–two plates with two identical API results.

13 / Septicaemia

Antibiotic susceptibility — Bacteria isolated from blood cultures are tested against a range of antibiotics using the disk diffusion method described previously. Since bacteria are first isolated in liquid media and a single species is usually present, the direct sensitivity method (see UTI, p. 33) can be used to provide an early guide to antibiotic therapy (Fig. 125). The inoculum may be too heavy or too light to obtain a readable result in some cases. In infective endocarditis, where bactericidal agents must be chosen from a small range of suitable agents, further tests are performed. These include the following.

Minimum inhibitory concentration (MIC)
The bacterial species isolated from the patient's blood culture is tested by exposure to a range of concentrations of antibiotic (e.g. penicillin) to find the lowest concentration at which growth is inhibited (Fig. 126).

Minimum bactericidal concentration (MBC)
After completion of the MIC test, aliquots are taken from each dilution and incubated to find the lowest concentration above which no bacteria can grow (Fig. 127). An MBC greatly in excess of the MIC is evidence that the organism is tolerant to the antibiotic tested.

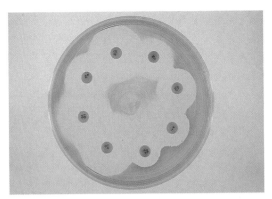

Fig. 125 Direct sensitivity test performed on *S. aureus* isolated from blood culture.

Fig. 126 Minimum inhibitory concentration performed by inoculating serial dilutions of antibiotic with test organism.

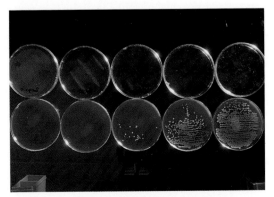

Fig. 127 Minimum bactericidal concentration, performed by subculturing from tubes in MIC test.

14 / Laboratory methods in infection control

One of the undesired consequences of modern medicine is hospital-acquired infection. The diagnostic laboratory has a crucial role to play investigating hospital outbreaks, sampling from individuals (patients and staff) and from the hospital environment.

Sampling from individuals

Either patients or members of staff may be reservoirs or become vehicles for agents of infectious disease. Confirmation of an infection hazard allows specific action to be taken which may prevent spread of disease. Common causes for concern and the recommended specimen include:

- diarrhoea—faeces
- hepatitis, AIDS—blood for serology
- pulmonary TB—sputum.

Asymptomatic or convalescent carriers of hospital pathogens (e.g. members of staff) may be sought after an outbreak or cluster of infections. Screening specimens may have to be collected from several individuals, or even large groups, in order to identify a probable source. Examples include:

- *S. aureus*—nose, groin, skin lesions (Fig. 128)
- *S. pyogenes*—throat
- *Salmonella* sp.—faeces.

The adequacy of hand hygiene can be assessed by hand impression plates, where staff are asked to press their hands onto agar plates before and after washing their hands (Fig. 129).

Sampling from the hospital environment

Environmental sampling may also help to identify specific infection hazards. The principal types of sampling are: surface swabs, slit sampling (known volume of air) and settle plates (used to culture airborne bacteria).

Fig. 128 Collection of screening cultures.

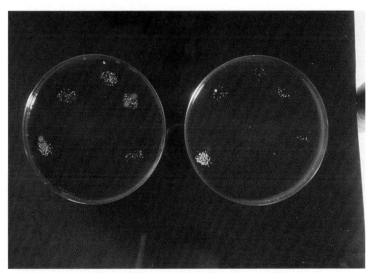

Fig. 129 Hand impression plates: before washing (left) and after washing (right).

14 / Laboratory methods in infection control

Establishing a common identity

If a single-source outbreak is to be attributed to a particular pathogen, it must be shown that all isolates of the suspected pathogen are identical. Establishing a common pathogen may confirm the presence of a single-source outbreak, while a lack of identity may provide evidence that no outbreak has taken place. Laboratory methods may also help trace the spread of infection through a population.

The basic requirement of any typing system is that a reaction should be obtained with every organism tested, with sufficient variety and reproducibility for categorization into a number of types or subtypes.

Methods

The tests most often used for epidemiological studies include:

- *Phage typing:* involves testing for susceptibility/resistance to lysis by bacteriophage (virus). It is used to type *S. aureus* isolates (Fig. 130).
- *Serotyping:* specific agglutination reactions used for *Salmonella* sp.
- *Pyocine typing:* inhibition of panel of *Pseudomonas* strains by *P. aeruginosa* strain isolated from patient or other source (Fig. 131).
- *Plasmid typing:* gel electrophoresis of extracted bacterial DNA. Identical banding patterns suggests common identity (Fig. 132). This is useful for many Gram-negative species.

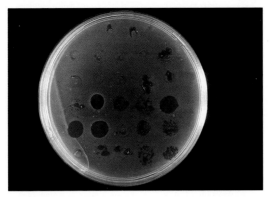

Fig. 130 Clear patches on lawn of test strain *S. aureus* indicate bacteriophage susceptibility pattern.

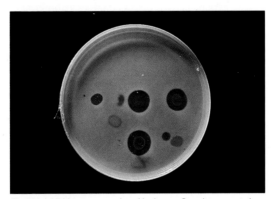

Fig. 131 Inhibition zones produced by known *Pseudomonas* strains in lawn of test *P. aeruginosa*.

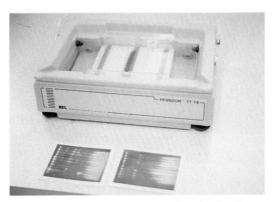

Fig. 132 Plasmid typing by gel electrophoresis: tank and result.

15 / **Virology**

Some virological tests can be performed in larger hospital laboratories, but most diagnostic virology is carried out in reference centres removed from the patient in question. For this reason, good communication with the diagnostic laboratory and a comprehension of the tests used are all the more important.

Specimens Specimens that may be required include the following:
- paired, clotted blood, for serology
- cerebrospinal fluid
- faeces
- vesicle fluid.

Blood should be collected for serological tests:
- during the acute phase of the illness
- after a 10 day interval (Fig. 133).

Specimens should not be tested serologically until the second sample has been collected, except after discussion with the laboratory.

Other virology specimens should be sent in viral transport medium (Fig. 134), which should be discarded if the contents has lost its red colour. This medium is not suitable for transporting specimens for fungal or bacterial culture.

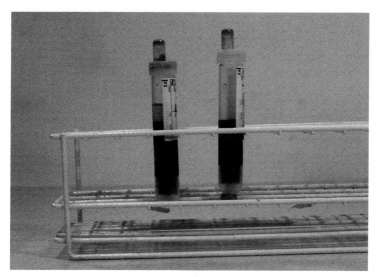

Fig. 133 Paired clotted blood samples for serological tests.

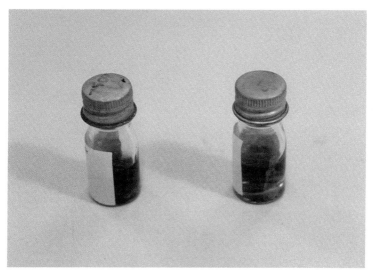

Fig. 134 Viral transport medium (left), with indicator colour change (right).

Microscopy Light microscopy is unsuitable for detection of viral particles because of their small size. Its main use is in examination of cell monolayers for viral inclusions and cytopathic effects (Fig. 139, p. 108).

The small size of viral particles means that the electron microscope has to be used to see viral structure in any detail. The small grids used to support specimens in the electron microscope permit examination of only very small specimens (Fig. 135). Uses include:

- *faeces*, e.g. for rotavirus (Fig. 136)
- *vesicle fluid*, for varicella zoster virus (Fig. 137)
- *immune electron microscopy*, uses antibody coated grids.

Immunological tests have replaced some electron microscopic work because they are less time consuming and give more consistent results.

Fig. 135 Electron microscope specimen grid.

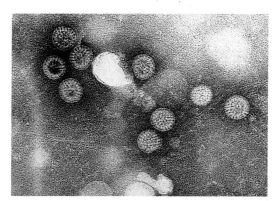

Fig. 136 Electron microphotograph of rotavirus particles.

Fig. 137 Electron microphotograph of varicella zoster virus.

15 / Virology

Culture Viruses are obligate intracellular parasites and cannot be cultivated without living cells. Monolayers of various cell types are used in the diagnostic laboratory to grow medically important viruses (Fig. 138).

Some virus types have effects that are specific to certain cell lines, so that they can be provisionally identified from their cytopathic effect (Fig. 139). It may be necessary to confirm their identity by comparing the results of virus inoculation on cell monolayers with and without the appropriate neutralizing antiserum.

Viruses of the herpes family can be grown in chick embryos. The size and time to appearance of pocks are characteristic of the different species.

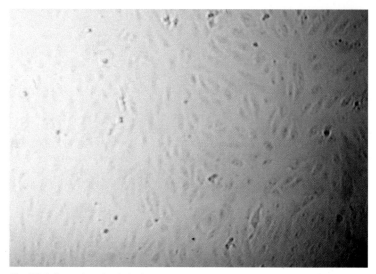

Fig. 138 Cell monolayer for viral culture.

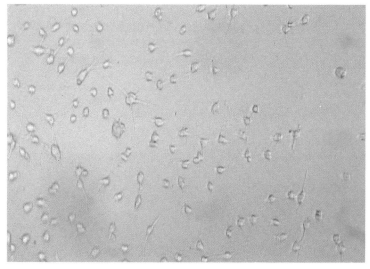

Fig. 139 Cell monolayer showing cytopathic effect following viral inoculation.

Serology   Serological techniques remain the most important group of diagnostic tests in clinical virology. Most serological tests rely on the detection of a rising antibody titre, in response to stimulation of the immune system by a specific viral antigen. In most tests, paired serum specimens are required with a 7–10 day interval between collection. It is usually not possible to comment on the significance of an antibody titre in a single serum specimen, particularly if it was collected during the acute phase.

Complement fixation test (CFT)
Antibody is detected when it binds complement and prevents activation of a complement-dependent haemolytic reaction in sensitized red cells. Though the test requires a series of controls and is time consuming, it is used for several antiviral antibody tests (Fig. 140).

Haemagglutination inhibition assay (HAI)
Antihaemagglutinin is detected when it binds with viral haemagglutinin, preventing agglutination of red cells (Fig. 141). Uses include diagnosis of rubella.

Enzyme-linked immunosorbent assay (ELISA)
ELISA tests have superseded many CFT and HAI based tests. The system may be designed to detect either antigen or antibody when it reacts with a captive antibody or antigen. This induces an enzymatic reaction which is seen as a colour change (Fig. 142).

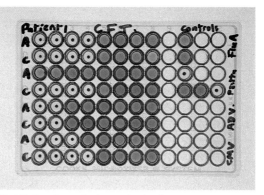

Fig. 140 Complement fixation test in microtitre tray.

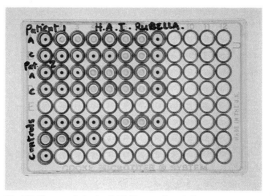

Fig. 141 Haemagglutination inhibition test for rubella.

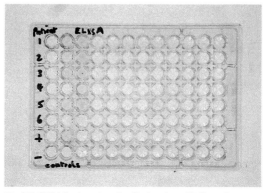

Fig. 142 ELISA for detection of antibody to hepatitis B e antigen.

Serology (cont) **Rubella radial haemolysis**
Sera from pregnant women are routinely screened for immunity to rubella virus using radial haemolysis. Serum is placed in wells in a gel containing sensitized red cells. There should be little or no haemolysis if protective antibodies are present (Fig. 143).

Immunoblotting
Certain immunoblotting techniques have came into use recently where highly specific results are required. Western blotting is used in some cases of AIDS, where the presence of antibodies to specific protein fractions of the virus relate to the course of the disease. A digest of HIV is separated by electrophoresis and loaded onto nitrocellulose strips. These are incubated with test serum and and indicator to give a banding pattern (Fig. 144).

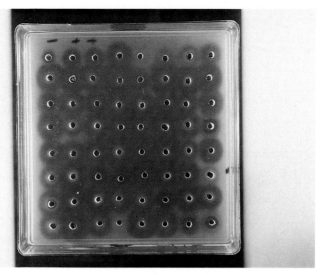

Fig. 143 Rubella radial haemolysis gel. The nonimmune control is at the top left corner.

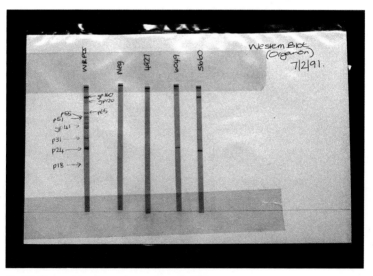

Fig. 144 Western blot of serum from patient with AIDS.

16 / **Mycology**

Specimens
Diagnostic mycology is concerned with the identification of medically important fungi. These include the dermatophytes and, increasingly, yeasts and moulds, some of which are now recognized as important agents of systemic infection in immunocompromised patients.

Superficial fungal infections are diagnosed by easily obtained specimens:

- skin—scrapings with a blunt blade (Fig. 145)
- nails—clippings
- hair—trimmings.

Invasive fungal infections require blood culture and/or soft tissue (for culture and histology). *Candida albicans* can be grown from standard aerobic blood culture bottles, which in some centres may be vented (Fig. 146).

Transport
Most fungal pathogens will tolerate transport to the laboratory.

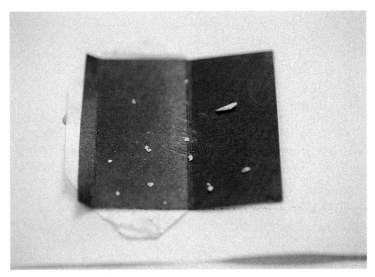

Fig. 145 Specimen of skin scrapings, sent in black card envelope.

Fig. 146 Vented blood culture bottle for fungal culture.

Microscopy Fungi other than yeasts are multicellular organisms with a complex morphology. Microscopic features of fungal species may help with identification.

Common microscopic preparation techniques are in KOH (for dermatophytes in skin specimens), and lactophenol cotton blue.

Dermatophytes

Three main genera cause skin and other superficial infections. These can usually be grown on mycological media, and are identified by their macroscopic and microscopic appearance.

- *Trichophyton* sp.—many microconidia, rare macroconidia (Fig. 147)
- *Epidermophyton* sp.—short macroconidia (Fig. 148)
- *Microsporum* sp.—spindle-shaped macroconidia (Fig. 149)

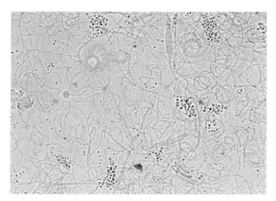

Fig. 147 *Trichophyton mentagrophytes* with many microconidia and spiral forms.

Fig. 148 *Epidermophyton floccosum* with short, thick-walled macroconidia.

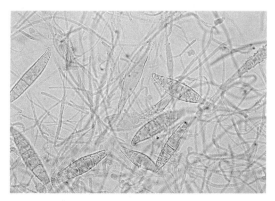

Fig. 149 *Microsporum canis* with spindle-shaped macroconidia.

16 / Mycology

The other two fungal genera commonly encountered in medical practice are *Candida* spp. and *Aspergillus* spp.

Candida species

These are yeasts—budding, unicellular organisms. They cause superficial infections of mucosal surfaces and systemic infections in immunocompromised patients.

Culture — The most commonly encountered member of the genus, *Candida albicans*, grows on Sabouraud's agar as white colonies (Fig. 150).

Confirmation — Its identity is usually confirmed with the germ tube test (Fig. 151).

Aspergillus species

These cause a number of conditions including lung disease and otitis externa.

Culture — Otitis externa is usually associated with *Aspergillus niger*, which produces black colonies on culture (Fig. 152).

Serology — In pulmonary aspergillosis, culture is rarely helpful, and aspergillus precipitins are of more use (particularly in allergic pulmonary aspergillosis and aspergilloma) (Fig. 153).

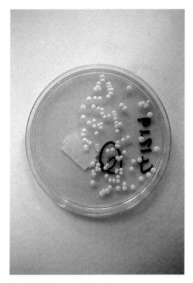

Fig. 150 Colonies of *C. albicans* on Sabouraud's agar plate.

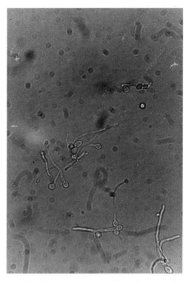

Fig. 151 Microphotograph of *C. albicans* pseudohyphae (germ tubes).

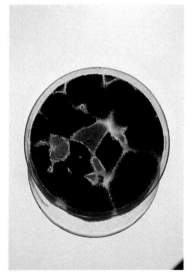

Fig. 152 Colonies of *Aspergillus niger*.

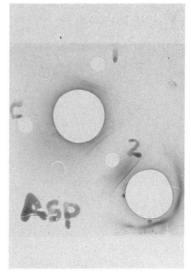

Fig. 153 Double diffusion plate test for aspergillus precipitins.

16 / Mycology

Therapeutic drug monitoring

The antifungal drug flucytosine is toxic in high concentrations. Serum concentrations are therefore measured in patients receiving this agent.

A plate bioassay is used which is similar to the antibacterial assay described above (p. 23). The serum concentration is calculated by comparing the inhibition zone diameter with those of a series of known standards tested at the same time (Fig. 154).

Antimicrobial susceptibility

Many fungi have a predictable susceptibility to commonly used antifungal agents, while the susceptibility of some of the newer therapeutic agents cannot yet be reliably tested. Susceptibility testing is therefore used less often in diagnostic mycology with the exception of invasive *C. albicans* infections, where a version of the disc diffusion test is used (Fig. 155).

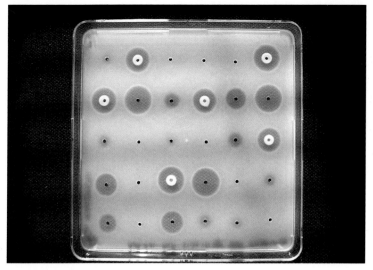

Fig. 154 Flucytosine plate bioassay with test and controls.

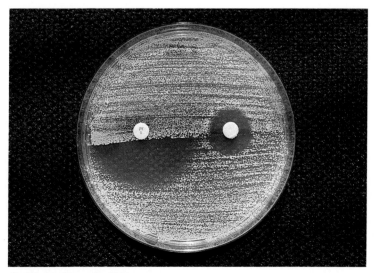

Fig. 155 *Candida* sp. disc diffusion test.

17 / **Parasitology**

Few agents of parasitic disease can be cultivated in the laboratory. In most cases, diagnosis is therefore based on microscopic techniques. Parasites with an enteric or blood stage to their life cycle can usually be identified in faecal or blood smears, respectively.

Specimens Fresh faecal specimens are best for parasitological work. Several may have to be collected to detect small numbers or irregular excretion of parasites. Blood smears should be prepared immediately after venesection. Vaginal *Trichomonas* sp. will survive collection and transport on a cotton swab.

Methods In many cases enteric parasites will not be present in sufficient numbers to detect by direct smear preparation. Concentration techniques are employed to remove the majority of faecal matter while retaining ova, cysts and vegetative forms. The most commonly used methods are:
- ether concentration (Fig. 156)
- barium sulphate flotation (Fig. 157).

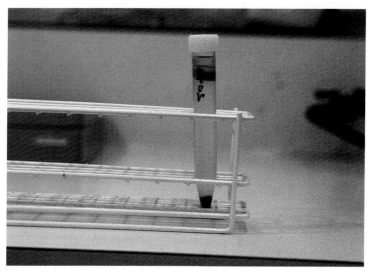

Fig. 156 Faeces separated into layers after ether treatment and centrifugation.

Fig. 157 Coverslip placed to collect parasites floating on top of barium sulphate.

Intestinal parasites

Protozoal infections

Giardiasis
The most common protozoan found in the UK is *Giardia lamblia (intestinalis)*. The cysts are difficult to see at microscopy and may only be passed in small quantities (Fig. 158). Several specimens should be examined before ruling out giardiasis. The trophozoites can also be recovered from duodenal aspirate or by using the swallowed string test.

Amoebiasis
Entamoeba histolytica, the cause of amoebic dysentery, can occasionally be seen on microscopic examination of still-warm stool specimens. However, diagnosis usually depends on the presence of cysts which must be differentiated from those of the non-pathogenic *Entamoeba coli* (Fig. 159). A serological test is also used in the diagnosis of invasive disease.

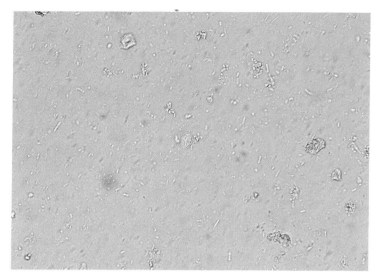

Fig. 158 Cysts of *Giardia lamblia*.

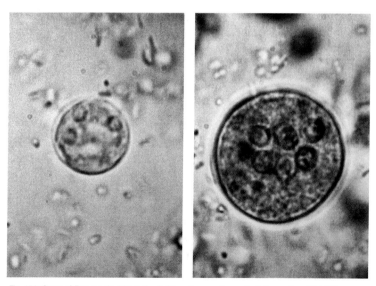

Fig. 159 Cysts of *Entamoeba histolytica* (left), and non-pathogenic *Entamoeba coli* (right).

Helminth infections

Enterobius vermicularis
Pinworm infection is diagnosed by applying adhesive tape to the perianal skin and examining for characteristic ova (Fig. 160).

Trichuris trichiura
Whipworm infection is diagnosed by microscopy of a faecal smear and recognition of typical ova (Fig. 161). Occasionally, the adult worm can be seen on the surface of the stool.

Ascaris lumbricoides
This infection is also diagnosed by faecal microscopy (Fig. 162). The adult worm may be passed per rectum.

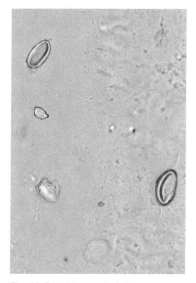

Fig. 160 *Enterobius vermicularis* ova on adhesive tape slide.

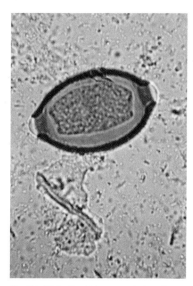

Fig. 161 *Trichuris trichiura* ovum in faecal smear.

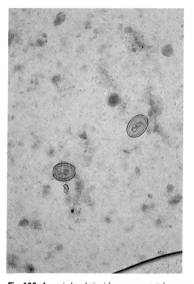

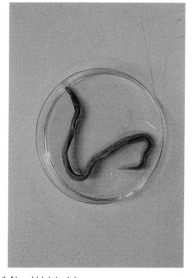

Fig. 162 *Ascaris lumbricoides* ovum on microscopy (left) and (right) adult worm.

Helminth infections (cont)

Hookworms
These organisms produce ova that can be recognized on faecal microscopy (Fig. 163), but the two types that infect humans (*Ancylostoma duodenale* and *Necator americanus*) cannot be distinguished by their microscopic appearance.

Tapeworms
These organisms also produce characteristic ova, recognizable on faecal microscopy (Fig. 164). The beef and pork tapeworms (*Taenia saginata* and *T. solium*, respectively) can only be distinguished by examination of body segments of the adult worm (proglottids) which may be passed per rectum.

Hydatid disease
This is also caused by a tapeworm, *Echinococcus granulosus* (or *E. multilocularis*). In this case, infection results in cyst formation in organs such as the lung, the liver and the brain. Microscopic examination of cyst fluid may reveal daughter cysts and hooklets.

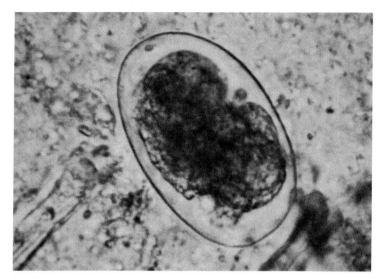

Fig. 163 Hookworm ovum.

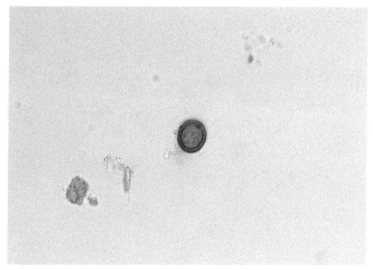

Fig. 164 Tapeworm ovum.

Trematode infections

Fasciola hepatica
This organism inhabits the bile ducts of infected individuals. Ova may be identified on examination of faeces or duodenal aspirate (Fig. 165).

Schistosomiasis
This is caused by three species which cause diseases with different epidemiologies and natural histories.

- *Schistosoma haematobium* may be passed in the urine. The ovum has a pointed terminal spine.
- *S. mansoni* localizes in the portal venous system of the large intestine, and is best diagnosed by microscopy of a rectal biopsy (Fig. 166). Microscopy of faeces is also worth attempting.
- *S. japonicum* often localizes around the small intestine. Diagnosis is by faecal microscopy (Fig. 167).

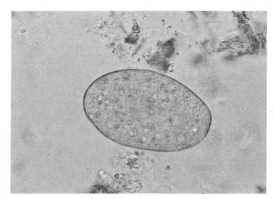

Fig. 165 Ovum of *Fasciola hepatica*.

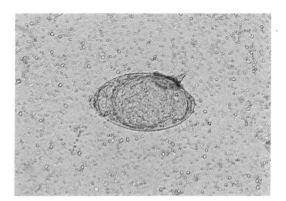

Fig. 166 *S. mansoni* ovum.

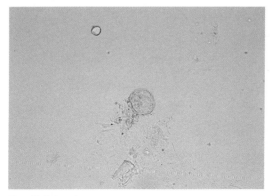

Fig. 167 *S. japonicum* ovum.

Bloodborne parasites

Many parasitic infections go through a stage of their life cycle in the bloodstream of their human host. In the bloodborne parasites, this stage is transmissible and may also be diagnosed by microscopic examination of a blood smear.

Clinical infections

Malaria
This is diagnosed by recognition of the ring forms, schizonts or gametocytes of *Plasmodium* sp. in either a blood smear (thin film, Fig. 168) or a lysed drop of blood (thick film, Fig. 168). The four species (*P. vivax*, *P. falciparum*, *P. ovale*, and *P. malariae*) can be differentiated by the characteristics of the three life cycle stages named above.

Trypanosomiasis
This infection can be diagnosed by microscopy of lymph node exudate or blood and, in the later stages of the disease, cerebrospinal fluid (Fig. 169).

Microfilariae
These can be found in the blood of patients with filarial diseases (Fig. 170). The different species can be distinguished by their morphological appearance.

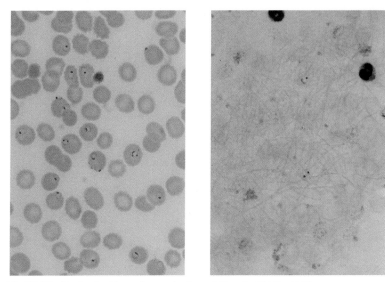

Fig. 168 Thin film showing *Plasmodium* sp. ring forms (left), and a thick film (right).

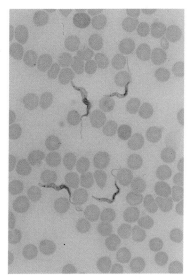

Fig. 169 *Trypanosoma* sp. in blood smear.

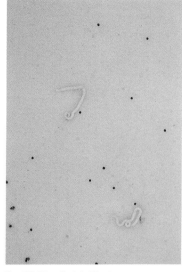

Fig. 170 Microfilaria in blood smear.

Index

Acid-fast bacilli, stains for, 59, 60
Actinomyces spp., culture, 85, 86
Agar plates, 9–12, *see also specific types of agar plate*
   in urinary tract infection, 29
Agglutination reactions, 19, 20, *see also* Coagglutination; Haemagglutination
   in meningitis, 71, 72
   in sexually transmitted disease, 41, 42
AIDS, HIV detection, 111
Amoebiasis, 123, 124
Anaerobic bacteria detected via gas liquid chromatography, 81, 82
Ancylostoma duodenale, 127
Antibiotics (antibacterial), 21–4
   in specimen, tests for presence, 31, 32
   susceptibility/resistance testing, 13, 21–2
     in gastrointestinal tract infection, 55, 56
     in meningitis, 77–8
     in respiratory tract infection, 65–6
     in septicaemia, 97–8
     in sexually transmitted disease, 43, 44
     in soft tissue infection, 83, 84, 86, 87–8
     in urinary tract infection, 33–4
   in therapy, monitoring, 23–4
Antifungal agents
   susceptibility, 119, 120
   in therapy, monitoring, 119
Ascaris lumbricoides, 125, 126
Aspergillus spp., identification and culture, 117, 118
Automation
   in antibiotic assays, 23, 24
   in blood culture screening for septicaemia, 91, 92

Babies, urinary tract infection specimens, 25
Bacilli, acid-fast, stains for, 59, 60
Bacteria, *see also specific disease/organism*
   in CSF, 69
   in urine (bacteriuria), 27, 29
Bacteriophage typing, 101
Bacteroides spp., culture, 85
Barium sulphate flotation, parasites collected via, 121, 122
Benchtop tests, 17–18
Bile solubility, *S. pneumoniae*, 63
Bioassay
   antibacterial agents, 23, 24
   antifungal agents, 119, 120
Biochemical tests, confirmatory, 19
   in gastrointestinal tract infection, 53, 54
   in respiratory tract infection, 63, 64
Blood
   CSF stained with, 67, 68
   culture
     in mycology, 113, 114
     in respiratory tract infection, 57
     screening, 91–2
     in septicaemia, 89, 90, 91–2
     in urinary tract infection, 31
   parasites in, 131–2
   poisoning, 89–98
   samples, in virology, 103, 104
Blood agars, 9, 10

Campylobacter spp.
   culture, 49
   microscopy, 47, 48
   preliminary identification, 51
Candida spp., 117
   albicans, culture, 39, 40, 117, 118
   antifungal susceptibility, 119, 120
   microscopy, 117, 118
Carbohydrate utilization tests
   in meningitis, 73, 74
   in sexually transmitted disease, 41, 42
Catalase test, 17, 18
Catheterized patients, urinary tract infection specimens, 25
Cerebrospinal fluid in meningitis, 67–78
   antibiotic susceptibility tests in cultures from, 77–8
   culturing, 73–6
   microscopy, 69–71
   rapid diagnosis from, 71, 72
   specimens, 67–8
Cervical smear, 37
Cervical swab, 35
Chancral exudate, 37
Children, urinary tract infection specimens, 25
Chlamydia trachomatis
   antibiotic susceptibility, 55, 56
   collection of specimens, 35
   culture, 39, 40
   microscopy, 37, 38
   transport of specimens, 37
Chocolate agars, 9, 10
CLED agar, 11, 12
Clostridium spp.
   difficile, toxin, 55, 56
   gas gangrene caused by, 81, 82
   perfringens, culture, 85
Coagglutination test, 19
   in meningitis, 71, 72
   in sexually transmitted disease, 41, 42
Coagulase test, 17, 18, 63
Complement fixation test for viruses, 109, 110
Confirmatory tests, 19–20
   in gastrointestinal tract infection, 53–4
   in respiratory tract infection, 63–4
   in sexually transmitted disease, 41–2
   in soft tissue infection, 85
Control of infection, laboratory methods in, 99–102
Corynebacterium diphtheriae, culture, 60, 62
Counterimmune electrophoresis of CSF, 71, 72

Index

Culture, 9–16
  of fungi, 117
  in gastrointestinal tract infection, 49–50
  incubation, conditions and apparatus, 15–16
  media used, see Media
  in meningitis, 73–6
  in respiratory tract infection, 57, 61–2
  in septicaemia, 95–6
  in sexually transmitted disease, 39
  in soft tissue infection, 83–6
  in urinary tract infection, 29–30, 31
  of viruses, 107–8
Cytopathic effects of viruses, 107, 108

DCA agar, 49, 50
Dermatophytes, 115, 116

Echinococcus spp.
  granulosus, 127
  multilocularis, 127
Electron microscopy
  techniques in, 3
  in virology, 105–6
Electrophoresis, counterimmune, of CSF, 71, 72
ELISA, 109, 110
Entamoeba spp.
  coli, 123, 124
  histolytica, 124
Enteric infection, see Gastrointestinal tract infection
Enterobacteriaceae
  antibiotic susceptibility, 21
  oxidase test, 17
Enterobium vermicularis, 125, 126
Enzyme(s), assays, 17–18
Enzyme-linked immunosorbent assay for viruses, 109, 110
Epidermophyton spp., 115, 116
Escherichia coli in meningitis, culture, 75
Ether concentration of parasites, 121, 122
Exudate
  chancral, 37
  inflammatory, see Pus

Faecal specimens
  in bacterial infection, 47, 48
  in parasitological infection, 121, 122
Fasciola hepatica, 129, 130
Fastidious organisms, recovery, 13
  in urinary tract infection, 31, 32
Filarial disease, 131
Flucytosine, therapeutic monitoring, 119, 120
Fluorescent treponemal antibody, absorbed (test), 45, 46
FTAabs test, 45, 46
Fungal infection, 113–20, see also specific diseases/fungi

Gangrene, gas, 81, 82
Gas liquid chromatography, anaerobes detected via, 81, 82
Gastrointestinal tract infection, 47–56, 123–30
  parasitological, 121, 123–30
German measles (rubella) virus, screening for immunity to, 111, 112
Giardiasis, 123, 124
Gonorrhoea, see *Neisseria* spp.
Gram stain, 5–6
  in gastrointestinal tract infection, 48
  in meningitis, 69, 70
  in respiratory tract infection, 59–60
  in septicaemia, 93, 94
  in sexually transmitted disease, 37
  in soft tissue infection, 81, 82
Gram-negative bacteria, appearance, 5, 6
Gram-positive bacteria, appearance, 5, 6
Growth factor tests, *Haemophilus* spp., 63, 64

Haemagglutination assay, *T. pallidum*, 45, 46
Haemagglutination inhibition assay for viruses, 109, 110
Haemolysis
  bacteria identified via, 9
  rubella immunity screening via radial, 111, 112
Haemophilus spp.
  confirmatory tests, 63, 64
  influenzae
    antibiotic susceptibility, 65, 66, 77, 78
    culture, 61, 62, 73
    growth factor tests, 63, 64
  parainfluenzae, growth factor tests, 63, 64
Hand hygiene, 99
Helminthic infection, 125–30
  gastrointestinal, 125–30
  haematological, 131
Herpes viruses, culture, 107, see also Varicella zoster virus
HIV, serological detection, 111
Hookworms, 127, 128
Hospital environment, infection control and sampling from, 99
Human immunodeficiency virus, serological detection, 111
Hydatid disease, 127, 128
Hygiene, hand, 99

Immunoblotting, viruses detected via, 111, 112
Immunofluorescence, *C. trachomatis* detection via, 37
Immunological tests
  for bacterial infection, confirmatory, 19
  for viral infection, 105
Incubation of cultures, conditions and apparatus, 15–16

134

Index

Infection control, laboratory methods in, 99–102, *see also specific sites/types of infection and specific (types of) pathogen*
Inflammatory exudate, *see* Pus
Inoculation medium, specimen, 13
Intestinal infection, *see* Gastrointestinal tract infection

[β]-Lactamase-production
  H. influenzae, 65, 66, 77
  N. gonorrhoeae, 43, 44
Lecithinase-producing clostridia, 85, 86
Leucocytes
  in CSF, 69
  in urine, *see* Pyuria
Light microscopy, techniques in, 3
Listeria monocytogenes in meningitis, culture, 75, 76
Lung infection, *see* Respiratory tract infection

MacConkey agar, 11, 12
Malaria, 131, 132
Media, culture, 9–14
  in gastrointestinal tract infection, 49, 50
  liquid, 13–14
  in meningitis, 73, 75
  non-selective, 9
  selective, 11
  in sexually transmitted disease, 39, 40
  in soft tissue infection, 83
  solid, 9–12
  in urinary tract infection, 29
Meningitis, 67–78
Microfilaria, 131, 132
Microscopy, 3–8
  in gastrointestinal tract infection, 47, 48
  in meningitis, 69, 71
  in mycology, 115–17
  in respiratory tract infection, 59–60
  in septicaemia, 93–4
  in sexually transmitted disease, 37, 38
  in urinary tract infection, 27–8
  in virology, 105–6
Microsporum spp., 115, 116
Minimum bactericidal concentration of antibiotic, 97, 98
Minimum inhibitory concentration of antibiotic, 97, 98
Mycobacteria spp.
  in meningitis, culture, 75
  microscopy, 7, 8
Mycology, 113–20

Nagler plate, 85, 86
Necator americanus, 127
Neisseria spp.
  gonorrhoeae
    antibiotic susceptibility, 43, 44
    collection of specimens, 35
    confirmatory tests, 41, 42
    culture, 11, 12, 39, 40
    microscopy, 37, 38
    transport of specimens, 37
  meningitidis
    antibiotic susceptibility, 77, 78
    culture-negative, 73
    microscopy, 69, 70
Nocardia spp., microscopy, 7, 8

Optochin sensitivity, *S. pneumoniae*, 63
Otitis externa, *Aspergillus niger* in, 117
Outbreaks, single-source, establishing, 101
Oxidase test, 17, 18

Parasitology, 121–32
Penicillin-resistant *N. gonorrhoeae*, 43, 44
Penicillinase-producing *N. gonorrhoeae*, 43, 44
Phage typing, 101
Pinworm, 125, 126
Plasmid typing
  in gastrointestinal tract infection, 55, 56
  in infection control, 101
Plasmodium spp., 131, 132
Proteus spp., preliminary identification, 51, 52
Protozoal infection
  gastrointestinal, 123–4
  haematological, 131
Pseudomonas spp.
  aeruginosa
    biochemical test, 63, 64
    inhibition of other *Pseudomonas* strains by, 101, 102
  antibiotic resistance, 33, 34
  oxidase test, 17, 18
Pulmonary infection, *see* Respiratory tract infection
Pus (inflammatory exudate)
  collection and transport, 79, 80
  microscopy, 81, 82
  rapid diagnosis from 81
Pyelonephritis, urinary tract infection specimens in, 25
Pyocine typing, 101
Pyuria, 27, 28
  sterile, 31

Quadferm, 41, 42

Respiratory tract infection, 57–66
  fungal, 117
Rotavirus, 105, 106
Rubella (German measles) virus, screening for immunity to, 111, 112

Sabouraud's agar, 39, 40
Salmonella spp.
  confirmatory tests, 53, 54
  preliminary identification, 51

Index

Sampling (specimen collection), 1, 2
  in gastrointestinal tract infection, 47
  infection control and, 99, 100
  in meningitis, 67
  in mycology, 113, 114
  in parasitology, 121, 122
  in respiratory tract infection, 57, 58
  in septicaemia, 89, 90
  in sexually transmitted disease, 35–6
  in soft tissue infection, 79, 80
  in urinary tract infection, 25
  in virology, 103, 104
Schistosoma spp.
  haematobium, 129
  japonicum, 129, 130
  mansonii, 129, 130
Screening of blood cultures in septicaemia, 91–2
Septicaemia, 89–98
Serological tests
  in gastrointestinal tract infection, 53, 54
  in infection control, 101
  in pulmonary aspergillosis, 117, 118
  in syphilis, 45–6
  in viral infection, 109–12
Sexually transmitted diseases, 35–46
Shigella spp.
  confirmatory tests, 53
  preliminary identification, 51
Single-source outbreaks, establishing, 101
Soft tissue infection, 79–88
Specimens, 1–2
  collection and transport, *see* Sampling; Transport
  gastrointestinal tract infection, 47
  inoculation medium, 13
  meningitis, 67–8
  mycological, 113, 114
  parasitological, 121
  respiratory tract infection, 57, 58
  septicaemia, 89, 90
  sexually transmitted diseases, 35–7
  soft tissue infection, 79, 80
  urinary tract infection, 25, 26
  virological, 99, 103, 104
Spore stain, 7, 8
Sputum specimen in respiratory tract infection, 57, 58
  culture, 61, 62
  microscopy, 59
Stains, 3–8
  in gastrointestinal tract infection, 48
  in meningitis, 69, 70
  in respiratory tract infection, 59–60
  in septicaemia, 93, 94
  in sexually transmitted disease, 37
  single-step, 3, 4
  in soft tissue infection, 81, 82
Staphylococcus spp.

aureus
  antibiotic susceptibility, 21
  coagulase test, 17, 18, 63
  catalase test, 17
  culture, 75, 76
Streptococci spp.
  catalase test, 17
  confirmatory tests, 63
  grouping, 63, 75
  meningitis caused by, 71, 72, 75
  pneumoniae
    antibiotic susceptibility, 65, 66
    bile solubility, 63
    culture, 61, 62, 71, 72
    optochin sensitivity, 63
  pyogenes, culture, 61, 62
  respiratory infection caused by, 61, 62, 63, 65, 66
Sugar utilization tests, *see* Carbohydrate utilization tests
Syphilis (*T. pallidum* infection), 45–6
  microscopy, 37
  serological tests, 45–6

Tapeworms, 127, 128
Throat swab in respiratory tract infection, 57, 58
  culture, 61, 62
Toxin production in gastrointestinal tract infection, 55, 56
TPHA (*Treponema pallidum* haemagglutination assay), 45, 46
Transport of specimens, 1
  in gastrointestinal tract infection, 47
  in meningitis, 67
  in mycology, 113
  in sexually transmitted disease, 37
  in soft tissue infection, 79, 80
  in urinary tract infection, 25, 26
  in virology, 103, 104
Trematode infection, 129–30
Treponema pallidum, *see* Syphilis; TPHA test
Trichomonas vaginalis, culture, 39
Trichophyton spp., 115, 116
Trichuris trichiura, 125, 126
Trypanosomiasis, 131, 132

Urethral smear, 37, 38
Urethral swab, 35, 36
Urinary tract infection, 25–34
Urine
  bacteria in, 27–9
  early morning, 31
  leucocytes in, *see* Pyuria

Varicella zoster virus, 137, 138
VCAT agar, 11, 12, 39, 40
VDRL (Venereal Disease Reference Laboratory) test, 45, 46

Virology, 103–12, *see also specific viruses and* Phage

Western blotting, HIV detected via, 111, 112
Wet films, 3
Whipworm, 125, 126

XLD agar, 11, 12, 49, 50

Yeasts, microscopy and culture, 117

Ziehl-Neelsen stain, 7, 8
   in respiratory tract infection, 59, 60